Air Pollution And The Politics of Control

Volume VI in MSS' series on Air Pollution

Papers by
Elliot Goldstein, Charles A. Cohen, Yves Alarie et al.

MSS Information Corporation
655 Madison Avenue, New York, N.Y. 10021

Library of Congress Cataloging in Publication Data
Main entry under title:

Air pollution and methods of control.

   1. Air-Pollution--Physiological effect--Addresses,
essays, lectures. 2. Air-Pollution--Addresses, essays,
lectures. I. Goldstein, E.   [DNLM: 1. Air
pollution--Prevention & control--Coll. wks. WA754 A294
1973]
RA576.A48   614.7'1'08   73-10421
ISBN 0-8422-7136-8

TABLE OF CONTENTS

CREDITS AND ACKNOWLEDGEMENTS

Alarie, Yves; Charles E. Ulrich; William M. Busey; Alex A. Krumm; and Harold N. MacFarland, "Long-term Continuous Exposure to Sulfur Dioxide in Cynomolgus Monkeys," *Archives of Environmental Health*, 1972, 24:115-128.

Allison, A.C., "Lysosomes and the Toxicity of Particulate Pollutants," *Archives of Internal Medicine*, 1971, 128:131-139.

Battigelli, Mario C.; David A. Fraser; and Homer Cole, "Microflora of the Respiratory Surface of Rodents Exposed to 'Inert' Particulates," *Archives of Internal Medicine*, 1971, 127:1103-1104.

Behrman, Richard E.; David E. Fisher; and John Paton, "Air Pollution in Nurseries: Correlation with a Decrease in Oxygen-carrying Capacity of Hemoglobin," *The Journal of Pediatrics*, 1971, 78:1050-1054.

Cohen, Charles A.; Arnold R. Hudson; Jack L. Clausen; and John H. Knelson, "Respiratory Symptoms, Spirometry, and Oxidant Air Pollution in Nonsmoking Adults," *American Review of Respiratory Disease*, 1972, 105:251-261.

Cota, Harold M., "Air Pollution Control Training in Colleges and Universities in the United States," *Air Pollution Control Association Journal*, 1971, 21:559-562.

Cotton, Raymond D., "The Federal Response to Air Pollution," *Archives of Environmental Health*, 1971, 23:243-244.

DuBois, Arthur B., "Establishment of 'Threshold' CO Exposure Levels," *Annals of the New York Academy of Sciences*, 1970, 174:425-428.

Goldstein, E., "Evaluation of the Role of Nitrogen Dioxide in the Development of Respiratory Diseases in Man," *California Medicine*, 1971, 115:21-27.

Parkinson, G.S., "Petroleum Fuels and Cleaner Air," *Chemistry in Britain*, 1971, 7:239-243.

Rigdon, R.H.; and Jack Neal, "Tumors in Mice Induced by Air Particulate Matter from a Petrochemical Industrial Area," *Texas Reports on Biology and Medicine*, 1971, 29:109-123.

Scorer, R.S., "New Attitudes to Air Pollution: The Technical Basis of Control," *Atmospheric Environment*, 1971, 5:903-934.

Shiffman, Morris A., "The Use of Standards in the Administration of Environmental Pollution Control Programs," *American Journal of Public Health*, 1970, 60:255-265.

Winkelstein, Warren, Jr.; and Michael L. Gay, "Suspended Particulate Air Pollution: Relationship to Mortality from Cirrhosis of the Liver," *Archives of Environmental Health*, 1971, 22:174-177.

PREFACE

In the United States air pollution reaches toxic thresholds in almost all urban areas. Its cost in terms of life and health is extreme. According to various governmental studies, the annual cost of damage resulting from polluted air is over $13 billion. If, for example, pollution could be cut by 50 percent, newborn babies would have an additional life expectancy of three to five years and deaths would be reduced by 4.5 percent. Air pollution is also a factor in the development of diseases ranging from respiratory disorders to cirrhosis of the liver, and polluted air intensifies allergic reactions. Serious long-term research and conscientious political action are needed if these ominous trends are to be reversed.

Volume VI in MSS' continuing series on air pollution presents current research on the toxicity of sulfur and nitrogen oxides, carbon monoxide, ozone, and particulate pollutants. Also included are studies of the political aspects of air pollution control, including papers on the problems of determining the cost factor of pollution, the training programs that produce specialists in pollution control, and descriptions of the efforts made by government and industry to combat air pollution.

Toxicity of Sulfur and Nitrogen Oxides, Carbon Monoxide, and Ozone

Evaluation of the Role

of Nitrogen Dioxide in

the Development of

Respiratory Diseases in Man

ELLIOT GOLDSTEIN, M.D.

IN RECENT YEARS, considerable evidence has accumulated indicating that nitrogen dioxide is a deleterious component of air pollutants.[1-5] Most of the evidence is descriptive and relates either human or animal exposure to some measurement of pulmonary dysfunction. It is implicit in these experiments that the measured abnormality eventually results in chronic pulmonary disease—for example, emphysema and bronchitis. Since these diseases affect approximately 4 percent of the population and their incidence is increasing,[6] this postulated sequence of events is of paramount practical importance. This report will review the available knowledge concerning the relationship of exposure to nitrogen dioxide and respiratory diseases.

Respiratory Effects of Nitrogen Dioxide In Man and Non-human Animals

The deleterious effects of air pollution are due to the interaction of many pollutants in addition to nitrogen dioxide, and hence a realistic evaluation of the significance of nitrogen dioxide requires data obtained within this context. Unfortunately, the complexities involved in investigating total atmospheres, have, in previous times, been insurmountable. As a result, epidemiological studies have usually contained measurements of only a few of the potential pollutants and experimental models have utilized artificial atmospheres in which nitrogen dioxide was the sole pollutant. It is anticipated that the engineering difficulties which have prevented measuring and simulating ambient atmospheres are near solution and in the future nitrogen dioxide will be studied in a more relevant manner.

Sources of Nitrogen Dioxide Contamination

Nitrogen dioxide enters the atmosphere as a by-product of natural gas combustion, following explosions, in industrial processes requiring the handling of nitric acid, and most importantly from burning petroleum in internal combustion engines. Effluent from gasoline engines contains nitric oxide, a minimally toxic pollutant which at ordinary temperatures is oxidized to the more toxic nitrogen dioxide. Although the complex photochemical reactions governing pollutant interactions are not well understood, sequences describing pollutant interaction have been delineated. According to Haagen-Smit and Wayne,[2] early morning automobile usage produces large quantities of nitric oxide and hydrocarbons. In the presence of sunlight, these products react, converting nitric oxide to nitrogen dioxide so that by mid-morning the atmosphere contains peak nitrogen dioxide and low nitric oxide levels. Subsequent irradiation of the nitrogen dioxide produces increasing concentrations of ozone during the afternoon and reduces the nitrogen dioxide to low levels. Late afternoon automobile traffic again produces large amounts of nitric oxide which reacts with the ozone to remove most of this pollutant, and then the nitric oxide continues to reaccumulate at a decreasing rate for the remainder of the night.

As the descriptive sequence indicates, in California the major source of nitrogen dioxide is combustion of automobile fuel; atmospheric concentrations increase in parallel with automobile ownership and use.[3] Thermal power plants are a second, major source of nitrogen dioxide. In heavily industrialized areas these utilities produce 25 to 50 percent of the total oxides of nitrogen.[3]

Although cigarette smoke does not contribute significantly to atmospheric pollution, the high concentrations of nitrogen dioxide which it contains may contribute to the enhanced incidence of respiratory disease of smokers.[7] The pathophysiology of this self-pollution is beyond the scope of this review and will not be considered further.

Information Pertaining to Chronic Exposures in Man

The significance to human health of presently encountered atmospheric levels of nitrogen dioxide is not known.[3,8] The older studies are difficult to evaluate because diagnostic criteria and data collection were inadequate.[9-11] Recent investigations involving healthy populations as well as patients with chronic respiratory diseases have yielded conflicting results. Shy et al compared neighboring communities in Chattanooga exposed to high and low concentrations of nitrogen dioxide.[4,5] They reported a decrease in ventilatory performance and an excess of respiratory illness among families exposed to the increased levels of nitrogen dioxide. Since the source of nitrogen dioxide pollution was a factory producing trinitrotoluene (TNT), other automobile-associated pollutants such as carbon monoxide, hydrocarbons and lead were not considered factors in this study. Spicer and Kerr on the other hand measured pulmonary vital capacity, total lung capacity, functional residual capacity and airway resistance at weekly intervals in 100 seminary students and did not find changes related to atmospheric concentrations of nitrogen dioxide.[12]

The effect of exposure to ambient levels of oxidant including nitrogen dioxide on pulmonary function in patients with chronic obstructive respiratory disease has also been investigated.[13-16] In two of these studies pulmonary function was evaluated during alternating periods in which patients breathed Los Angeles smog for a week and then decontaminated air for a week, and in both studies improvement in respiratory function was noted when the patients breathed decontaminated air.[13,14] Objective measurements showing improved ventilatory function were the 3-second timed vital capacity, maximal breathing capacity, and a reduced residual lung volume.[13] It is important to note that particulates and ozone as well as nitrogen dioxide were removed by the purification procedures, and hence these results may not pertain directly to nitrogen dioxide. Rokaw and Massey observed a group of 25 patients with chronic obstructive disease for 18 months and did not find a correlation between air pollution levels including nitrogen dioxide and subjective or objective measurements of pulmonary function.[15] Burrows and coworkers in a larger but less precise investigation also failed to find significant relationships between air pollutant levels and pulmonary function in patients with chronic obstructive pulmonary disease.[16]

Acute Exposures in Man

Definitive evidence demonstrating that exposure to nitrogen dioxide can be deleterious to respiratory function comes from two sources. Acute exposure of humans to high levels of nitrogen dioxide invariably results in respiratory disease.[17-22] In experimental animals exposed to elevated levels of atmospheric nitrogen dioxide pathologic changes resembling emphysema develop and susceptibility to bacterial infection is enhanced.[23-34]

The average level of nitrogen dioxide in atmosphere of smog-ridden areas of California is 0.25 ppm.[2] Air pollution surveys indicate a maximal concentration of 3.5 ppm of nitrogen dioxide.[23] Acute exposure to higher levels of nitrogen dioxide is an uncommon occupational hazard of workers manufacturing nitric acid,[11] of farmers (exposed to silage—silo-fillers' disease),[17,18] and of electric arc workers.[19] A range of adverse effects correlating with the degree of exposure have been described. Eye and nasal irritation occurs after exposure to 15 ppm of nitrogen dioxide. Pulmonary discomfort is noted at levels of 25 ppm and bronchiolitis with focal pneumonitis occurs after exposures of 25 to 75 ppm of nitrogen dioxide. The duration of these exposures was put at less than one hour. Comparably short periods of exposure to 150-200 ppm causes fatal pulmonary fibrosis. Higher exposures are associated with acute pulmonary edema, sometimes death.[1,17] These studies demonstrate conclusively that elevated concentrations of nitrogen dioxide are extremely toxic to human respiratory tissues.

Volunteer studies further support the hypothesis that acute exposure to concentrations of nitrogen dioxide above ambient, will impair pulmonary function. Abe[20] observed an increased expiratory and inspiratory flow resistance in healthy adults exposed for 10 minutes to 4 to 5 ppm of nitrogen dioxide. Higher levels of exposure (50 ppm) for 1 minute have been shown to cause significant nasal irritation and pulmonary discomfort.[21]

Animal Experiments

The heightened contemporary interest in the potential toxicity of nitrogen dioxide has led to many investigations with experimental animals. These experiments have certain inherent deficiencies. In most cases, the animal model is similar but not completely analogous to the human. As an example, the respiratory anatomy of rats (the animal most frequently used in studies of nitrogen dioxide induced emphysema) differs from

14

that of humans in: not having interlobular septae; having fewer generations of airways; utilizing distal bronchioles for respiration rather than alveoli; pulmonary vasculature.[35] The epithelium of the tracheobronchial tree of rats also differs from man in that it has more large, mucus-secreting glands lining the trachea and fewer along the bronchi.[36] There are also important anatomic differences between humans and rabbits, mice, guinea pigs, monkeys, and dogs.[35] These distinctions in anatomy may explain why identical concentrations of nitrogen dioxide cause diverse pathologic disturbances and make interpretation that is relevant to disease in man exceedingly difficult.

Pathologic Abnormalities Following Exposure to Nitrogen Dioxide

Most studies concerned with the pathology of prolonged exposure to nitrogen dioxide have utilized rodents.[28-30,32,37,38] In rats exposed to 10, 12.5, or 25 ppm of nitrogen dioxide for three or more months the thoracic cavities become larger, dorsal kyphosis develops and the animals have an inflated appearance. Microscopically there is distension of alveolar ducts, dilation of alveoli and hyperplasia of bronchiolar epithelium. Alveolar septa are occasionally missing, but destruction of parenchyma is unusual.[30,37] These pathologic features are similar but not identical with those of human emphysema. A critical difference is the absence of alveolar necrosis. Destructive bullous lesions are the *sine qua non* of emphysema;[39] bullae are absent in rodent models. Also, it should be noted that the lesions mentioned above do not develop in rats exposed to lower concentrations of nitrogen dioxide (0.8 to 2.0 ppm) for their entire lifetime. The lungs from these animals are grossly normal; microscopic examination shows only minor degrees of ciliary loss, epithelial hypertrophy and "cytoplasmic blebbing."[32,38] These animals live out a normal life span and die of diseases unrelated to nitrogen dioxide.[37]

Mice are more susceptible to the toxic effects of nitrogen dioxide than rats. Continuous exposure to 0.5 ppm of nitrogen dioxide for three months causes loss of cilia, alveolar cell disruption and obstruction to respiratory bronchioles.[31] Exposures of longer duration cause more severe changes and pneumonitis.[31] These pathologic abnormalities are, however, unlike the changes of emphysema in man.

Haydon et al exposed rabbits continuously to atmospheres containing 8 to 12 ppm of nitrogen dioxide for 3 to 4 months and reported destructive changes in alveolar walls and abnormal enlargement of the distal air spaces.[27] These findings closely approximate the emphysematous lesions observed in humans. Unfortunately, there are no reports of data obtained from rabbits exposed to lower concentrations of nitrogen dioxide. It should also be noted that other investigators have failed to find emphysematous changes in rabbits exposed for two hours per day to 15 to 25 ppm of nitrogen dioxide for periods up to two years.[10] A multifocal type of emphysema has been induced in guinea pigs following three weeks of exposure for two hours daily to 22 ppm of nitrogen dioxide.[11] Since these levels are considerably above ambient, the relevance of this model to human disease states is doubtful. The hamster appears to be particularly resistant to the toxic effects of nitrogen dioxide. Kleinerman and Cowdry exposed hamsters to 45 to 55 ppm of nitrogen dioxide for 21 to 23 hours daily for ten weeks and did not find emphysematous changes.[10] The dog is also resistant to the toxic effects of nitrogen dioxide. Wagner et al exposed dogs to 5.0 ppm of nitrogen dioxide for 15 to 18 months and did not find differences between the lungs of treated and control animals.[34] These distinctive results have been confirmed by other investigators.[28,33] Investigations with monkeys are currently in progress but have not yet been reported.[42] Horses, the laboratory animals whose pulmonary anatomy most closely approximates that of man, have, apparently, not been studied.

Effect of Nitrogen Dioxide on Pulmonary Resistance to Infection

Ehrlich and coworkers in a series of experiments have shown that acute and chronic exposure to relatively low levels of nitrogen dioxide depresses pulmonary resistance to infection in mice.[23,24,31] The experimental method consisted of exposing animals to atmospheres of fixed nitrogen dioxide concentration before infection with aerosols of virulent Klebsiella sp. The technique has been sufficiently standardized to predict that 25 to 50 percent of infected controls will die of pneumonia caused by the Klebsiella sp. within 14 days. In a few experiments, bacterial clearance rates were obtained by removing lungs at various time intervals after infection and determining the concentrations of bacteria.

In the acute experiments, mice were exposed for two hours to concentrations of nitrogen dioxide (1.5 to 25 ppm) before infection. Significant increases in mortality occurred in the animals exposed to levels above 3.5 ppm. Deaths did not occur in uninfected mice exposed to identical concentrations of nitrogen dioxide.[23] Further studies with this murine model showed that the adverse effect of nitrogen dioxide is transient. Animals infected with virulent Klebsiella microorganisms 27 hours after exposure to 5, 15 or 25 ppm of nitrogen dioxide did not have an increased mortality when compared with controls.

These investigators also showed that continuous exposures to levels of nitrogen dioxide only slightly above ambient (0.5 ppm or more for three months) depressed murine resistance to pulmonary infection.[24]

Studies of pulmonary bacterial clearance mechanisms using the previously mentioned bacterial clearance technique have demonstrated that the enhancement of murine susceptibility to infection by K. pneumoniae was caused by diminished pulmonary antibacterial activity. Mice which are exposed to nitrogen dioxide and then challenged with aerosols of K. pneumoniae are unable to kill

17

the inhaled bacteria as well as untreated controls. This decrease in bacterial clearance rate is directly proportional to the intensity of exposure to nitrogen dioxide.

Experiments in which hamsters were exposed to nitrogen dioxide and then infected with aerosols of Klebsiella pneumoniae also demonstrated impairment in pulmonary clearance mechanisms, but only at very high concentrations of nitrogen dioxide.[23] This relative increase in resistance to the effect of nitrogen dioxide exposure was attributed to both a diminished virulence of Klebsiella pneumoniae for hamsters and an increased resistance to the adverse effects of nitrogen dioxide.

A few experiments with monkeys have been performed.[25,43] A two hour exposure to 10 to 50 ppm of nitrogen dioxide depressed resistance to aerosol challenge with Klebsiella pneumoniae.[43] Exposure to 10 ppm of nitrogen dioxide for one month or 5.0 ppm for two months also resulted in an enhanced susceptibility to infection. The latter data are preliminary since only a few monkeys were studied; thus, one of four monkeys died following exposure to 10 ppm of nitrogen dioxide and two of seven died at the 5.0 ppm level.[25]

The effect of nitrogen dioxide in combination with other pollutants has been studied by Coffin and coworkers.[44] In these studies, exhaust from an automobile was photochemically treated and conveyed into exposure chambers. Mice within the chambers were exposed to the auto exhaust for four hours and then infected with aerosols of Streptococcus. An enhanced mortality was noted in animals exposed to 25 ppm of carbon monoxide and 0.15 ppm of oxidant. Since nitrogen dioxide forms a significant percentage of exhaust oxidant, it is likely that in this situation it contributed to the adverse effect.

In one report, exposure to nitrogen dioxide did not cause a decrease in pulmonary anti-bacterial activity.[45] Buckley and Loosli exposed germfree and conventional mice to 38 ppm of nitrogen dioxide for six weeks. At the end of this period the

18

animals were infected with aerosols of S. aureus, and bacterial clearance rates were determined during the next five days. Although the rates of bacterial clearance for the germfree and conventional mice differed, neither group was affected by exposure to nitrogen dioxide. The investigators interpreted their data as showing that nitrogen dioxide did not effect bacterial clearance rates. These studies may be criticized on two counts. First, only three animals were studied at each time period—too few for statistical analysis. Second, S. aureus is not pathogenic for mice. Within 24 hours, 99 percent are removed and hence significant differences in clearance that might have occurred within the initial 24-hour period would have been overlooked.

It should be noted that it is very unlikely that the differences in the data of these studies were due to the kinds of microorganisms that were studied (that is, the pathogen K. pneumoniae and the non-pathogen S. aureus). Previous investigations with similar murine models have clearly shown that differences in bacterial virulence cause quantitative but not qualitative changes in clearance rates.[46,47]

Effect of Nitrogen Dioxide on Alveolar Macrophage Function

From the previously cited in vivo studies, exposure to nitrogen dioxide appears to inhibit alveolar macrophage function.[23,25,31,43] A few studies have been reported in which macrophages were exposed to nitrogen dioxide in vitro.[48,49] According to the data from one investigation, macrophages are killed by exposure to extremely high levels of nitrogen dioxide, 176 ppm.[48] Myrvik and Evans, in a more elegant study, exposed alveolar macrophages from rabbits to 50 ppm of nitrogen dioxide and demonstrated a significant reduction in phagocytic function with a concomitant suppression of cellular energy pathways.[49] Before evaluating data obtained from in vitro exposures to nitrogen dioxide, it should be recognized that the effective concentration of nitrogen dioxide in the fluid phase is undoubtedly much lower than in the air phase due to the instability of nitrogen dioxide in water. Hence the atmospheric concentrations used cannot be equated with the levels reported in *in vivo* experiments.[49]

Effect of Nitrogen Dioxide On Mucociliary Function

Few studies relating exposure to nitrogen dioxide to mucociliary function have been performed.[50-52] According to the data obtained, ciliary activity is inhibited by exposure to nitrogen dioxide and this defect results in a decreased rate of particle removal. Certain deficiencies in the laboratory model deserve emphasis. Clearance rates were measured in isolated tracheal segments from either rabbits or rats. An isolated segment is divorced from neuromuscular control, blood supply and the effects of deglutition. Moreover, the rate that particles move along the trachea may not be representative of the rate of function of the entire tracheobronchial tree. Finally, the nitrogen dioxide exposures were short-term and at levels considerably above ambient.

Recently, an improved method of studying mucociliary function has been reported by Spritzer and coworkers.[53] A tight-fitting tube is inserted surgically into the esophagus of the rat and then attached, via the stomach, to an external collecting bottle. Radiolabeled particles are either aerosolized or injected intratracheally and the rate of entrance into the collection bottle is measured. Curves of particle removal for "normal" rats have been reported. Further experiments with this model should allow a more accurate determination of the effect of nitrogen dioxide on mucociliary function.

Effect of Nitrogen Dioxide On Immune Mechanisms

In preliminary studies nitrogen dioxide appears

to have an effect upon immune reactions. Matsumura recently demonstrated that exposure of guinea pigs sensitized to egg albumin to 70 ppm of nitrogen dioxide for 30 minutes enhances their susceptibility to systemic anaphylaxis when challenged with aerosols of egg albumin.[54] Exposure to lower concentrations of nitrogen dioxide (40 ppm) causes an increase in the severity of the dyspneic symptoms.[55] Circulating antibody reactive with pulmonary tissue has also been found in guinea pigs that were exposed for as long as one year to 5.0 to 15.0 ppm of nitrogen dioxide.[56] Since these experiments·were conducted at pollutant exposure levels much above ambient, further investigations will be necessary before these data can be related to human instances of pulmonary disease.

Nitrogen Dioxide as a Biological Oxidant

The biochemical mechanisms by which nitrogen dioxide causes cellular dysfunction are in the initial stages of investigation. Since nitrogen dioxide and ozone are similar, it may be that some of the toxic effects of nitrogen dioxide result from biological oxidation to form free radicals.[1] Thomas and associates have presented evidence to support this important hypothesis.[57] These investigators showed an increase in lipoperoxidation of lung lipids in rats exposed to 1.0 ppm of nitrogen dioxide four hours daily for six days. Of practical significance is the additional finding that pre-treatment with high levels of anti-oxidant (10 mg of vitamin E per day) was partially effective in preventing the lipid peroxidative changes induced by nitrogen dioxide. Although it is hazardous to extrapolate from data obtained in laboratory models of infection to instances of human disease, nevertheless these experimental results raise the intriguing possibility that the ingestion of anti-oxidants might prevent some of the deleterious consequences accruing from exposure to nitrogen dioxide.

21

Conclusion

The studies that have been cited document severe pulmonary disease in individuals exposed acutely to very high concentrations of nitrogen dioxide. However, these concentrations are much above ambient and their relevance to daily environmental exposures is minimal. Since concern about the potential danger of nitrogen dioxide is quite recent, and the postulated disease processes are chronic, definitive information relating pulmonary disease to exposure to nitrogen dioxide under actual, ambient conditions is not available.

There are epidemiological data, however, that support the idea that respiratory impairment may occur in healthy populations and in patients with chronic obstructive respiratory disease following exposure to atmospheric levels of nitrogen dioxide. Although not conclusive, this evidence is sufficient to justify extensive epidemiological investigations designed to determine the significance to human health of exposure to ambient concentrations of nitrogen dioxide. Animal models have served as a valuable means for determining the pathophysiological effects of exposure to pollutants. These studies have shown that exposure to nitrogen dioxide in concentrations that exceed those ordinarily encountered results in pathologic abnormality of the bronchi and alveoli and an enhanced pulmonary susceptibility to bacterial infection.

At present, extrapolation from these data to man is hazardous since animal models do not truly reflect the environmental-host relationships of human exposure. However, as newer techniques are developed, quantitative data delineating the pathophysiological effects, if any, of ambient exposures to nitrogen dioxide should become available and allow insight into the biological consequences of chronic exposure of man to nitrogen dioxide.

REFERENCES

1. Stokinger HE, Coffin DL: Biologic effects of air pollutants, chap 13. *In* Stern AC: Air Pollution and Its Effects. New York, Academic Press, 1968, pp 445-546

2. Haagen-Smit AJ, Wayne LG: Atmospheric reactions and scavenging processes, chap 6. *In* Stern AC: Air Pollution and Its Effects. New York, Academic Press, 1968, pp 149-186

3. Cooper WC, Tabershaw IR: The oxides of nitrogen in air pollution. Report to the California Department of Public Health, July 1967, pp 1-107

4. Shy CM, Creason JP, Pearlman ME, et al: The Chattanooga school children study—I. Methods, description of pollutant exposure, and results of ventilatory function testing. J Air Pollut Contr Assoc 20:539-545, 1970

5. Shy CM, Creason JP, Pearlman, et al: The Chattanooga school children study—II. Incidence of acute respiratory illness. J Air Pollut Contr Assoc 20:582-588, 1970

6. Goldsmith JR, Fallat RJ: Prospects for the prevention of disabling pulmonary disease. Milbank Mem Fund Quart 47:235-249, 1969

7. Smoking and Health—Report of the Advisory Committee to the Surgeon General of the Public Health Service. US Dept of HEW, US Public Health Service Publication No 1103, 1964

8. Wolkonsky PM: Pulmonary effects of air pollution—Current research. Arch Environ Health 19:586-592, 1969

9. Vigdortschik NA, Andreeva EC, Mattuswitch IL, et al: Symptomatology of chronic poisoning with oxides of nitrogen. J Ind Hyg Toxicol 19:469-473, 1937

10. Vigliani EC, Zurlo N: Erfalrungen der clinica del Levaro mit einegen maximalen arbeitsplatzkonsentration (MAK) von Industriegiffen. Arch Gewerbepath U Gewerbehyg 13:528-534, 1955

11. Patty FA (Ed): Industrial Hygiene and Toxicology—Vol 2. New York, Wiley Interscience, 1962, pp 920-923

12. Spicer WS, Kerr HD: Effects of environment on respiratory function—III. Weekly studies on young male adults. Arch Environ Health 21:635-642, 1970

13. Motley HL, Smart RH, Leftwich CI: Effect of polluted Los Angeles air (smog) on lung volume measurements. JAMA 171:1469-1477, 1959

14. Remmers JE, Balchum OJ: Effects of Los Angeles urban pollution upon respiratory function of emphysematous patients. Read before the 58th Annual Meeting of the Air Pollution Control Association, Toronto, June 22, 1965

15. Rokaw SN, Massey F: Air pollution and chronic respiratory disease. Amer Rev Resp Dis 86:703-704, 1962

16. Burrows B, Kellogg AL, Buskey J: Relationship of symptoms of chronic bronchitis and emphysema to weather and air pollution. Arch Environ Health 16:406-413, 1968

17. Lowry T, Schuman LM: Silo-Filler's disease syndrome caused by nitrogen dioxide. JAMA 162:153-160, 1956

18. Grayson RR: Silage gas poisoning: Nitrogen dioxide pneumonia, a new disease in agricultural workers. Ann Intern Med 45:393-408, 1956

19. Norwood WD, Wisehart DE, Earl LA, et al: Nitrogen dioxide poisoning due to metal cutting with oxyacetylene torch. J Occup Med 8:301-306, 1966

20. Abe M: Effects of mixed NO_2-SO_2 gas on human pulmonary functions. Bull Tokyo Med Dent Univ 14:415-433, 1967

21. Meyers FH, Hine CH: Some experiences with nitrogen dioxide in animals and man. Fifth Air Pollution Med Research Conference, Los Angeles, 1961

22. Tse RL, Bockman AA: Nitrogen dioxide toxicity: Report of 4 cases in firemen. JAMA 212:1341-1344, 1970

23. Ehrlich R: Effect of nitrogen dioxide on resistance to respiratory infection. Bact Rev 50:604-614, 1966

24. Ehrlich R, Henry MC: Chronic toxicity of nitrogen dioxide—I. Effect on resistance to bacterial pneumonia. Arch Environ Health 17:860-865, 1968

25. Henry MC, Findlay J, Spangler J, et al: Chronic toxicity of NO_2 in squirrel monkeys—III. Effect on resistance to bacterial and viral infection. Arch Environ Health 20:566-570, 1970

26. Carson TR: The responses of animals inhaling nitrogen dioxide for single short-term exposures. Amer Industr Hgy Assoc J 23:457-462, 1962

27. Haydon GB, Davidson JT, Lillington GA, et al: Nitrogen dioxide-induced emphysema in rabbits. Amer Rev Resp Dis 95:797-805, 1967.

28. Steadman BL, Jones RA, Rector DE, et al: Effects on experimental animals of long-term continuous inhalation of nitrogen dioxide. Toxic Appl Pharmacol 9:160-170, 1966

29. Freeman G, Stephens RJ, Crane SC, et al: Lesion of the lung in rats continuously exposed to two parts per million of nitrogen dioxide. Arch Environ Health 17:181-192, 1968

30. Freeman G, Crane SC, Stephens RJ, et al: Pathogenesis of the nitrogen dioxide-induced lesion in the rat lung: A review and presentation of new observations. Amer Rev Resp Dis 98:429-443, 1968

31. Blair WH, Henry MC, Ehrlich R: Chronic toxicity of nitrogen dioxide—II. Effects on histopathology of lung tissue. Arch Environ Health 18:186-192, 1969

32. Freeman G, Furiosi NJ, Haydon GB: Effects of continuous exposure of 0.8 ppm NO_2 on respiration of rats. Arch Environ Health 13:454-456, 1966

33. Vaughan TR Jr, Jennelle LF, Trent TR: Long term exposure to low levels of air pollutants—Effects on pulmonary function in the beagle. Arch Environ Health 19:45-50, 1969

34. Wagner WD: Experimental study of threshold limit of NO_2. Arch Environ Health 10:455-465, 1965

35. Tyler WS, McLaughlin RF Jr, Canada RO: Structural analogues of the respiratory system. Arch Environ Health 14:62-69, 1967

36. Roe FJC: The relevance and value of studies of lung tumours in laboratory animals, In Research on Cancer of the Human Lung. In Lung Tumours in Animals, Severi I. (Ed). Perugia, June 1966, pp 111-126

37. Freeman G, Crane SC, Stephens RS, et al: Environmental factors in emphysema and a model system with NO_2. Yale J Biol Med 40:566-574, 1968

38. Haydon GB, Freeman G, Furiosi NJ: Covert pathogenesis of NO_2 induced emphysema in the rat. Arch Environ Health 11:776-783, 1965

39. American Thoracic Society: Definitions and classifications of chronic bronchitis, asthma and pulmonary emphysema. Amer Rev Resp Dis 85:762-764, 1962

40. Kleinerman J, Cowdrey CR: The effect of continuous high level nitrogen dioxide on hamsters. Yale J Biol Med 40:579-590, 1968

41. Gross P, deTreville RTP, Babyak BS, et al: Experimental emphysema—Effect of chronic nitrogen dioxide exposure and papain on normal and pneumoconiotic lungs. Arch Environ Health 16:51-58, 1968

42. Crane SC: Personal communication

43. Henry MC, Ehrlich R, Blair WH: Effect of nitrogen dioxide on resistance of squirrel monkeys to Klebsiella pneumoniae infection. Arch Environ Health 18:580-587, 1969

44. Cohn DK, Blommer EJ: Acute toxicity of irradiated auto exhaust. Its indication by enhancement of mortality from streptococcal pneumonia. Arch Environ Health 15:36-38, 1967

45. Buckley RD, Loosli CG: Effects of nitrogen dioxide inhalation on germfree mouse lung. Arch Environ Health 18:588-595, 1969

46. Green, L, Green GM: Differential suppression of pulmonary antibacterial activity as the mechanism of selection of a pathogen in mixed bacterial infection of the lung. Amer Rev Resp Dis 98:819-824, 1968

47. Goldstein E, Green GM: Alteration of the pathogenicity of Pasteurella pneumotropica for the murine lung caused by changes in pulmonary antibacterial activity. J Bact 93:1651-1656, 1967

48. Weissbecker L, Carpenter RD, Luchsinger PC, et al: In vitro alveolar macrophage viability—Effect of gases. Arch Environ Health 18:756-759, 1969

49. Myrvik QN, Evans DG: Metabolic and immunologic activities of alveolar macrophages. Arch Environ Health 14:92-96, 1967

50. Battigelli MC, Hengstenberg F, Mannela RJ, et al: Mucociliary activity. Arch Environ Health 12:460-466, 1966

51. Kenster CJ, Battista SP: Components of cigarette smoke with ciliary-depressant activity—Their selective removal by filters containing activated charcoal granules. New Eng J Med 269:1161-1166, 1963

52. Dalhamn T, Sjoholm J: Studies on SO_2, NO_2 and NH_3. Effect on ciliary activity in rabbit tracheal of single in vitro exposure and resorption in rabbit nasal cavity. Acta Physiol Scand 58:287-291, 1963

53. Spritzer AA, Watson JA, Auld JA: Mucociliary clearance rates—Deposition and clearance in the tracheobronchial tree of rats. Arch Environ Health 15:39-47, 1967

54. Matsumura Y: The effects of ozone, nitrogen dioxide, and sulfur dioxide on the experimentally induced allergic respiratory disorder in guinea pigs—I. The effect on sensitization with albumin through the airway. Amer Rev Resp Dis 102:430-443, 1970

24

55. Matsumura Y: The effects of ozone, nitrogen dioxide, and sulfur dioxide on the experimentally induced allergic disorder in guinea pigs—III. The effect on the occurrence of dyspneic attacks. Amer Rev Resp Dis 102:444-447, 1970

56. Balchum OJ, Buckley RD, Sherwin R, et al: Nitrogen dioxide inhalation and lung antibodies. Arch Environ Health 10:274-277, 1965

57. Thomas HV, Mueller PK, Lyman RL: Lipoperoxidation of lung lipids in rats exposed to nitrogen dioxide. Science 159:532-534, 1968

Respiratory Symptoms, Spirometry, and Oxidant Air Pollution in Nonsmoking Adults

CHARLES A. COHEN, ARNOLD R. HUDSON, JACK L. CLAUSEN, and JOHN H. KNELSON

Introduction

Urban air pollution can be divided into two major categories. The reducing type, such as that found in London, consists mainly of carbonaceous, particulate matter and sulfur dioxide. The oxidizing form, such as that found in the Los Angeles area, consists mainly of the primary pollutants (hydrocarbons and the oxides of nitrogen) and photochemical reaction pollutants (ozone, nitrogen dioxide, aldehydes, peroxyacetyl nitrate, and other organic nitrates). Convincing evidence has been presented implicating the reducing type of air pollution as a factor contributing to increased mortality in persons with pulmonary and cardiac disease, in both the United Kingdom and the United States (1, 2). Epidemiologic studies correlating high prevalences of chronic lung disease with prolonged exposure to the reducing type of air pollution (1, 3) suggest that this type of environmental insult might be an important factor in the pathogenesis of these diseases, although it is not nearly as important as

26

cigarette smoking.

In the few epidemiologic studies conducted, no consistent association between excess mortality (4, 5) and the prevalence (4, 6) or incidence (7) of chronic lung disease with values of oxidant air pollution was shown; however, the deleterious effects of this type of environmental pollution on plants, synthetic fabrics and dyes, and rubber are well documented (8). In addition, acute and subacute human exposure to oxidant values somewhat greater than the usual ambient ones resulted in reversible decrements in ventilatory function (9–11).

In the present study the hypothesis, that chronic exposure to large peak values of oxidizing air pollution in the absence of cigarette smoking is associated with an increased prevalence of chronic lung disease and irreversible changes in ventilatory function, was tested. Socioeconomic status (1, 2, 12), meteorologic conditions (13), occupation (14), and ethnic background (15) were considered, because correlation between these variables and respiratory disease has been shown.

Seventh-Day Adventists were selected for this study for the following reasons: (1) very few of them smoke; (2) the church roster provides an accessible census; (3) previous studies of the incidence of respiratory disease in Seventh-Day Adventists in California could be used for comparison and (4) they are interested in health and would be likely to participate.

The San Gabriel Valley was chosen as one study area because of its large peak values of oxidant pollution; the San Diego area was selected as the other because it has lower peak oxidant values although many other major pollutants and meteorologic conditions are similar to those in the San Gabriel Valley. The study design required minimizing the effects of acute exposure to oxidants in order to maximize the possibility of detecting irreversible chronic effects. Large oxidant concentrations usually occur during the summer and fall seasons (16); therefore, the winter season was selected because the concentrations of oxidants in both areas were expected to be minimal and about equal during that time.

Materials and Methods

Selection of study population: Six churches in the San Gabriel Valley and four churches in the San Diego area were selected because of proximity to central testing locations. Active church members were eligible for study if they were white, English-speaking, between the ages of 45 years and 64 years (inclusive), had not smoked in the last 20 years, and had a lifetime history of smoking of less than one pack per year. Inactive church members, defined as those who had not attended church in the previous six months for reasons other than bad health, and those whose phone numbers or addresses were not available, were excluded. Each of the participating churches provided a list of those who would probably meet the study criteria. A church member validated the eligibility status of each person on the preliminary list by telephone call. Any member not found after five calls was asked by postcard to list a time and place where he could be found. Each volunteer was mailed an appointment notice during January 1970 for an interview by a physician and for tests of ventilatory function. Those who did not indicate acceptance of the appointment by postcard were contacted by telephone to arrange a more suitable appointment. After the study was completed, all eligible members who refused to volunteer or who were unable to meet their appointments were asked to complete the respiratory symptom questionnaire in a telephone interview. The respiratory questionnaire was a modified combination of the Questionnaire on Respiratory Symptoms approved by the British Medical Research Council Committee on Research into Chronic Bronchitis and the London School of Hygiene Cardiovascular Questionnaire (17) and was administered by the same interviewer in both areas. Social class was estimated by the Hollingshead method using data on education and occupation (18). Residents of San Diego who had lived in Los Angeles or Riverside for more than five years since 1945 or any time in the previous five years were excluded from data analysis.

Concentrations of pollutants: For four weeks

during a summer month (September 1969) and one month during the study period (January 1970), the following atmospheric measurements were made by National Air Pollution Control Administration personnel at four sampling sites in each area: (*1*) total suspended particulates, (*2*) respirable suspended particulates, (*3*) sulfur dioxide, (*4*) oxidants, (*5*) nitrates, and (*6*) sulfates. In addition, concentrations of oxidants during each day were obtained from the Pasadena and Azusa stations of the Los Angeles County Air Pollution Control District and the downtown San Diego station of the San Diego County Air Pollution Control District.

Air pollution values measured in the San Diego and San Gabriel Valley area by the County Air Pollution Control Districts since 1959 were reviewed to compile data representative of chronic exposure (19). The Pasadena and Azusa air sampling stations were selected as sites representative of the San Gabriel Valley; the methods of analysis were similar for all three stations.

Pulmonary function testing: For spirometry and determinations of maximal expiratory flow-volume (MEFV) curves, the subjects were seated and wore nose clips; each test was done by the same technician using the same apparatus in both areas. Vital capacity was measured twice using a 13.5 liter Stead-Wells[3] spirometer; another replicate was made when the first two did not agree. The procedure used was as follows: after a few tidal volumes, the subject exhaled slowly to his expiratory reserve volume, then inhaled to total lung capacity; the drum was speeded up to 32 mm per sec, and the subject then executed a forced expiration. The starting time for the forced expiration was determined by extrapolation of the down slope of the spirogram to the maximal inspiratory line. Inspiratory vital capacity (IVC), forced expiratory volume at one second (FEV_1), and maximal forced expiratory flow from 200 ml to 1,200 ml ($FEF_{200-1,200}$ or MEF) were determined according to the recommendations of the American College of Chest Physicians (20). The maximal values of these parameters were selected for analysis. The third quartile times (TQT) were determined by measuring the time between the

[3] Warren E. Collins Co., Boston, Massachusetts.

TABLE 1

TOTAL NUMBER OF PERSONS QUALIFIED FOR THE STUDY AND
PER CENT PARTICIPATION BY LOCATION

Location	Total No. of Members	Total Qualified	Per Cent Who Completed Questionnaire (Including Telephone Interview)	Per Cent Who Performed Spirometry
San Gabriel Valley	1,917	197	95.9	67.5
San Diego	1,956	244	98.3	82.4
Total	3,873	441	97.2	75.7

expiration of 50 per cent and 75 per cent of the IVC. Inspiratory vital capacity was used instead of expiratory vital capacity (EVC) for the following reasons: (1) although in normal subjects the forced expiratory vital capacity (FVC) is usually very similar to the EVC done slowly, in some subjects with obstructive lung disease the FVC gives significantly lower values for the VC than the EVC done slowly. Measurement of both forced expiratory flow (FEF) and EVC, however, requires two separate VC maneuvers; both a slow IVC and FEF can be measured with one maneuver (complete expiration, maximal inspiration, followed by a forced expiration). This reduction in the number of maneuvers represents a considerable saving of time when many subjects are studied in a day. (2) Inspiratory vital capacity and EVC were not significantly different in a study by Mills (21).

Maximal expiratory flow-volume curves were obtained by using a 14-liter Med-Science low resistance, low-inertia, piston-type spirometer (Model 465)[4] that uses two separate transducers for the direct measurement of flow and volume, and a rapid-response Hewlett-Packard XY recorder (Model AFM).[5] All curves were replicated, and a third or fourth curve was obtained when the flows in the last two thirds of the FEV could not be superimposed. For each subject, the parameters were measured from the curve with the highest peak expiratory flow (PEF). Flows were measured at volumes when 50 per cent of IVC ($\dot{V}_{50\%}$) and 75 per cent of IVC

[4] Med-Science Electronics, St. Louis, Missouri.
[5] Hewlett-Packard Co., Boston, Massachusetts.

TABLE 2

PER CENT DISTRIBUTION OF SOCIAL CLASS* OF SUBJECTS
COMPLETING VENTILATORY FUNCTION TESTS

Location	Social Class					Total No. of Subjects
	1	2	3	4	5	
San Gabriel Valley	19.9	20.6	21.3	33.1	5.1	136
San Diego	22.7	7.7	34.3	25.6	9.7	207

*Estimated by Hollingshead method (16) using data on education and occupation. Class 1 contains persons in the professions; Class 5, unskilled laborers with no college education.

($V_{75\%}$) were expired. A time constant (TC) in seconds was determined by calculating the reciprocal of the slope of the line drawn between $V_{50\%}$ and $\dot{V}_{75\%}$. All volumes were corrected to body temperature and pressure, saturated with water vapor (BTPS).

Results

Study populations: The total membership of the ten churches selected was 3,873, of which 441 members qualified for the study; 97.2 per cent of those eligible completed the questionnaire (table 1). Ventilatory function was tested on a greater proportion of subjects from the San Diego area (82.4 per cent) than from the San Gabriel Valley (67.5 per cent) whereas the questionnaire was completed for approximately the same percentage (98.3 per cent and 95.9 per cent, respectively) in both areas. The ratio of women to men answering the questionnaire was 2.5 : 1 in San Diego and 2.2 : 1 in the San Gabriel Valley. A similar ratio was present in those who underwent pulmonary function testing. The two groups did not differ significantly in age or social class (table 2). Of those tested, 89.5 per cent had lived in the area of their present residence for five years or longer; 78.1 per cent, for ten years or longer.

The over-all prevalence of chronic bronchitis, as defined by cough productive of phlegm on most days during three or more

TABLE 3

RESPIRATORY SYMPTOMS BY PER CENT OF PARTICIPANTS IN SPECIFIED AGE, SEX, AND AREA GROUPS

| | Ages 45–54 years | | | | Ages 55–64 years | | | | All Ages — Both Areas | | |
| | Men | | Women | | Men | | Women | | | | |
	San Gabriel Valley n=33	San Diego n=42	San Gabriel Valley n=66	San Diego n=89	San Gabriel Valley n=21	San Diego n=33	San Gabriel Valley n=69	San Diego n=76	Men n=129	Women n=300	Both Sexes n=429
No cough or phlegm	84.8	85.7	84.8	87.6	80.9	87.8	87.0	85.2	85.2	86.0	85.8
Nonpersistent cough or phlegm	15.2 (5)	11.9 (5)	12.1 (8)	10.1 (9)	14.2 (3)	6.0 (2)	11.6 (8)	11.8 (9)	11.6 (14)	11.3	11.4
Nonpersistent cough and phlegm	0	0	1.5 (1)	2.2 (2)	4.7 (1)	3.0 (1)	0	0	1.6 (2)	1.0 (3)	1.2 (5)
Persistent* cough and phlegm (chronic bronchitis)	0	2.3 (1)	1.5 (1)	0	0	3.0 (1)	1.4 (1)	3.9 (3)	1.6 (2)	1.7 (5)	1.7 (7)

*Persistent means on most days during three months a year or more for at least two consecutive years.
†Figures in parentheses refer to number of subjects.

months a year for at least two consecutive years, was only 1.7 per cent (table 3). The prevalence rate for cough with phlegm lasting less than three months of the year was 1.2 per cent. There were no significant or consistent differences between the two areas for any of the respiratory symptom complexes. Also, there were no significant differences between those who appeared at the test locations and those who later completed the questionnaire during a telephone interview.

Pollutant concentrations: As expected, the "acute peak oxidant exposure," namely, the daily maximal hourly average during the study, was small in both areas, i.e., 0.12 ppm in the San Gabriel Valley and 0.07 ppm in San Diego. Likewise, the suspended particulate value was small and about equal in both areas, averaging 75 μg per m^3. Of the total suspended particulates, the respirable fraction (diameter of less than 10 μ) was approximately 63 per cent in both areas. Sulfur dioxide (SO_2) is not one of the major air pollutants in southern California. In September 1969, none of the measurements of SO_2 in either area was greater than 0.01 ppm, which approaches the minimal detectable concentration for this method. In January 1970 during the study period, mean SO_2 concentrations were 0.0032 ppm in San Diego and 0.0045 ppm in the San Gabriel Valley. There was, therefore, no evidence of any acute air pollution episode during the study period in either area.

The years 1963 to 1967 were selected as a period of time representative of chronic exposure for which complete data were available. The concentrations of oxidants during this period are summarized in table 4. Although 75 per cent of the hourly values in San Diego were greater than those in San Gabriel, these differences were small and were at very low concentrations. Both areas had almost identical annual mean values for all hourly oxidant concentrations; however, the means of the daily maximal hourly concentrations in San Gabriel were twice

TABLE 4

SUMMARY OF OXIDANT CONCENTRATIONS (PPM) IN THE SAN GABRIEL VALLEY AND SAN DIEGO, 1963–1967

Area	Cumulative Per Cent Frequency Distributions of Hourly Average Concentrations										Arithmetic Mean of Hourly Concentrations	Arithmetic Mean of Daily Maximal Hourly Concentrations	Daily Maximal Hourly Average Equal to or Exceeding 0.15 ppm % of days	Highest Daily Maximal Hourly Average	Highest Peak Measurement
	10	30	50	60	70	80	90	95	99						
Azusa	0.01	0.01	0.02	0.03	0.04	0.08	0.14	0.20	0.30	0.049	0.150	45.7%	0.65	0.73
Pasadena	0.01	0.01	0.02	0.02	0.04	0.07	0.12	0.18	0.27	0.044	0.138	43.8%	0.46	0.67
San Diego	0	0.02	0.03	0.04	0.05	0.06	0.08	0.10	0.15	0.038	0.074	6.1%	0.80	1.01

the value measured in San Diego, and the number of days when the daily maximal hourly average equaled or exceeded 0.15 ppm (a concentration at which many persons experience symptoms of irritation of eyes, nose, and throat) was more than seven times greater in San Gabriel than in San Diego. The annual mean concentrations of the other major gaseous pollutants are summarized in table 5; concentrations were generally greater in San Gabriel although peak values were occasionally higher in San Diego. The average total suspended particulate values during this five-year period, measured by the high volume method, were larger in Los Angeles (121 μg per m^3) than in San Diego (73 μg per m^3).

Pulmonary function tests: The spirometric and flow-volume parameters in the two areas were tested by analysis of variance for each of the two sexes with the values of age, height, and socioeconomic status as covariates (table 6). Each value for each of the covariates was adjusted by a linear regression equation. There were no significant differences between the two areas for any of the pulmonary function parameters tested, and all of the mean values were well within values considered normal (table 7).

When the analysis was limited to those who lived in an area for at least five years (299 members) or to those in residence for more than ten years (261 members), the same pattern was demonstrated. Spirometric values were studied for two other subgroups: the 42 with current or resolving respiratory tract infections and the 5 who gave a history of occupational exposure to dust for more than one year. They did not differ significantly from the total population.

Discussion

The percentage of the membership of the churches that was eligible for this study was considerably lower in the San Gabriel Valley area than in San Diego. There were probably two reasons for this. (*1*) One large church in the San Gabriel Valley had a predominance

35

TABLE 5
SUMMARY OF MAJOR AIR POLLUTANT CONCENTRATIONS IN THE SAN GABRIEL VALLEY AND SAN DIEGO DURING 1963–1967.

Pollutant	Area	Cumulative Per Cent Frequency Distribution of Hourly Average Concentrations (ppm)				Arithmetic Mean of Hourly Concentrations (ppm)
		50%	90%	95%	99%	
Carbon Monoxide	Azusa	9	12	13	15	9.2
	Pasadena	10	16	19	25	11.2
	San Diego	2	6	9	15	2.9
Nitric oxide	Azusa	0.01	0.04	0.06	0.11	0.019
	Pasadena	0.02	0.13	0.19	0.32	0.050
	San Diego	0.01	0.10	0.18	0.38	0.035
Nitrogen dioxide	Azusa	0.04	0.09	0.11	0.16	0.045
	Pasadena	0.05	0.11	0.13	0.20	0.057
	San Diego	0.01	0.06	0.07	0.12	0.023
Hydrocarbons	Azusa*	2	4	5	6	2.7
	Pasadena†	3	5	6	8	3.3
	San Diego**	3	5	7	10	3.2
Sulfur dioxide	Azusa	–	–	–	–	–
	Pasadena	0.01	0.03	0.03	0.05	0.015
	San Diego††	–	–	–	–	–
Suspended particulates (sampling period=24 hours), μg/m³	Azusa***	–	–	–	–	–
	Pasadena***	122	168	–	–	124
	San Diego	68	110	–	–	78

* 1963–1964.
† 1967.
** 1965–1967.
†† Continuous measurements of sulfur dioxide were not made in San Diego after 1959. Intermittent measurements since then consistently measured values at or near zero.
*** 1964–1966.

TABLE 6

COMPARISON OF PULMONARY FUNCTION
MEASUREMENTS FROM THE SAN GABRIEL
VALLEY AND SAN DIEGO BY ANALYSIS OF
VARIANCE

Parameter	Men		Women	
	F value	P	F value	P
Inspiratory vital capacity	0.41	0.524	0.65	0.42
FEV_1	0.66	0.42	0.88	0.35
Maximal expiratory flow	3.20	0.08	2.01	0.16
Third quartile time	0.02	0.88	0.55	0.46
Peak expiratory flow	0.95	0.33	0.20	0.66
Expiration of 50% inspiratory vital capacity	1.26	0.27	0.09	0.77
Expiration of 75% insiratory vital capacity	0.94	0.34	0.66	0.42
Time constant	1.42	0.24	1.93	0.17
FEV_1/Inspiratory vital capacity	0.00	0.96	0.55	0.46

of young adults and retirees; (2) The directory of one large church in the San Gabriel Valley was somewhat outdated, and consequently, a larger percentage of the membership could not be reached by telephone and mail surveys. The preponderance of females eligible for study was expected; a previous census of all Seventh Day Adventists in California showed the ratio of female to male church members in the 45-year to 64-year age range was approximately 2 : 1 (22). One reason for the slightly higher sex ratio documented among study volunteers was the disqualification of more men than women because of cigarette smoking. San Gabriel Valley and San Diego volunteers were comparable with respect to socioeconomic status, age, sex, and height. Because a previous study documented the validity of smoking histories obtained from Seventh Day Adventists in California, there was very little likelihood of smoking differences between cities (23). Although those interviewed but not tested had more complaints referable to non-respiratory illness than those interviewed and tested, analyses of the questionnaire showed that the former were not significantly different from the latter with respect to respiratory symptoms, and therefore the ventilatory function tests were assumed to be representative for the entire group from each area.

The prevalence rate for chronic bronchitis was unexpectedly low. A much higher prevalence rate for persistent cough (23 per cent) was observed by Wynder among 239 California male Seventh Day Adventist lifetime nonsmokers (24); however, in that study, clearing of the throat was recorded as a cough. Furthermore, the writers admitted that most of the male Seventh Day Adventists who reported persistent cough were confusing throat clearing with true bronchial cough. The observed prevalence rate for chronic bronchitis of 1.77 per cent was much lower than that observed by Ferris and associates in nonsmokers in Berlin, New Hampshire (13.8 per cent and 9.47 per cent in men and women, respectively); whether this difference was due to inherent differences in the populations or the environment cannot be ascertained with the available data.

In the present study, the extremely low prevalence of chronic bronchitis restricted the value of statistical comparison by sex or age. The greater prevalence of chronic bronchitis for women in both areas is nonetheless worth noting. Previous mortality studies have shown that Seventh Day Adventists in California are unusual in that mortality from chronic lung diseases and carcinoma of the lung was approximately equal for men and women (25, 26). As expected, the prevalence of chronic bronchitis increased with age; no person less than 50 years old reported symptoms of chronic bronchitis.

Many studies have emphasized the effect of socioeconomic status on ventilatory function tests; however, in this study no effect of socioeconomic status on ventilatory function was observed despite the wide ranges of occupational and educational levels of the participants. On the other hand, there probably was not a socioeconomic group in this study comparable to the lowest socioeconomic groups in other studies; even the least privileged volunteers in this study were gainfully employed. Furthermore, it was the investigators' opinion that this population was relatively homogeneous in terms of diet, adequacy of housing, and medical care, which might have accounted for the homogeneity seen in the results of pulmonary function tests.

Air pollution agencies in California, particularly in the Los Angeles area, have for many years been collecting some of the most sophisticated and complete pollutant data in this country. Aerometry is a relatively costly and complex undertaking, and in this study that correlated health effects with air pollution, it is important to review a few of the problems inherent in using the aerometry data currently available for a retrospective study. One important limitation was the relatively small number of sampling stations that were in operation in the past decade in the areas studied. Measurements of various air pollutants can vary considerably over an area of a few square miles because of meteorologic conditions and localized emission sources. Caution is necessary in interpreting data from one or two stations, even though they are well placed, when trying to correlate air pollution with health effects in people living in a large area such as San Diego or the San Gabriel Valley.

The pollutant concentrations, however, were measured at four sampling locations within each area, which were selected as more representative of the places of residence of the subjects than the established sampling stations. It was discovered that most of the pollutant values were quite uniform in the areas studied in San Diego; in the San Gabriel Valley, the concentrations of nitrates and particulates were somewhat larger at the sampling locations closer to the mountains,

TABLE 7

MEAN (± SD) VALUES FOR PULMONARY FUNCTION PARAMETERS BY AREA AND SEX

Spirometry	San Gabriel Valley		San Diego	
	Men n=41	Women n=90	Men n=64	Women n=135
Inspiratory vital capacity, liter	4.47 ± 0.72	3.08 ± 0.55	4.41 ± 0.66	3.09 ± 0.48
FEV_1, liter	3.54 ± 0.51	2.44 ± 0.42	3.49 ± 0.52	2.43 ± 0.45
Maximal expiratory flow, liter/sec	8.60 ± 1.94	5.25 ± 1.20	7.81 ± 1.43	5.10 ± 1.41
Third quartile time	0.239 ± 0.075	0.235 ± 0.073	0.239 ± 0.074	0.244 ± 0.111
Flow volume				
Inspiratory vital capacity, liter	4.46 ± 0.74	3.03 ± 0.57	4.33 ± 0.70	3.02 ± 0.47
Peak expiratory flow, liter/sec	8.26 ± 0.96	5.83 ± 1.03	7.92 ± 1.12	5.82 ± 1.09
$\dot{V}_{50}$, liter/sec*	4.86 ± 1.55	3.38 ± 0.83	4.49 ± 1.32	3.37 ± 1.03
$\dot{V}_{75}$, liter/sec*	1.51 ± 0.52	1.00 ± 0.35	1.54 ± 0.52	1.07 ± 0.45
Time constant sec^{-1}	0.384 ± 0.169	0.341 ± 0.102	0.408 ± 0.149	0.374 ± 0.192
Age, years	52.07 ± 5.15	54.01 ± 5.75	54.19 ± 5.39	53.48 ± 5.58
Height, inches	69.46 ± 3.24	63.27 ± 2.50	68.89 ± 2.74	63.55 ± 2.26
Social class	2.54 ± 1.27	2.97 ± 1.19	2.58 ± 1.33	3.01 ± 1.22

*See table 6 for definitions of symbols.

but the oxidants and sulfates were quite evenly distributed. It was concluded that the established sampling stations in these areas were measuring concentrations representative of the pollutant values to which the subjects were exposed.

Another limitation in analyzing aerometry data retrospectively is that the comparison of pollutant data obtained by different agencies using different analytic techniques can be a major problem. Fortunately, essentially the same methods of analysis for the pollutants were used in all three stations from which data were analyzed.

As illustrated in table 4, it is important to appreciate that many different average concentrations of oxidants can be computed from the same pollutant data simply by varying the averaging time. For example, using a one-hour averaging time, as was done in this study, the data showed that values in the San Gabriel Valley exceeded those in San Diego approximately only 25 per cent of the time. If a 12-hour averaging time was used, however, San Gabriel Valley had higher values approximately 65 per cent of the time. It should also be emphasized that oxidant pollution is not a single entity, but a complex combination of several individual pollutants such as ozone, nitrogen dioxide, and photochemically active hydrocarbons. So-called oxidant concentrations are measures of the oxidizing power of the atmosphere that contains these individual components, ozone being the main component. It might be more meaningful to look at the concentrations of these components separately rather than collectively; with the limitations of current technology, however, this would be an enormously complicated and expensive task. Thus, it is obvious that researchers in human health effects of air pollution will benefit from further advances in aerometric methods. Indeed, more meaningful population studies might depend on this.

Winkelstein and associates (2) have shown greater mortality from chronic respiratory disease (after correcting for economic status) that was related to residence in areas with total suspended particulate values of 100 μg to 135 μg per m^3 per 24 hours. The SO$_2$ concentrations, although not great, were larger than those found in either of the areas in this study. There is experimental evidence that synergism exists between the effects of SO$_2$ and particulates on airway resistance, and it is possible that the low concentrations of SO$_2$ in the San Gabriel Valley explain why the mean particulate values of 121 μg per m^3 in this area were not demonstrably associated with a greater prevalence of symptoms of respiratory disease or poorer pulmonary function tests.

Spirometric measurements, although widely used as an index of severity in patients with lung disease, have certain limitations. Recent experimental (27) and clinical (28) evidences indicate that obstruction in bronchi and bronchioles can be quite extensive before measurable changes in the total airway resistance or flow occur. For relatively insoluble gaseous pollutants, there is theoretic (29) as well as experimental (30) and clinical (31) evidence that the primary toxic effect of inhalation occurs in the small bronchi and bronchioles, precisely the area whose function is not directly measurable by conventional testing. Hopefully, further development of new approaches in measuring airway resistance, such as the frequency dependence of compliance (32), will permit noninvasive measurements to be made of this "quiet zone" (33) of the lung.

Chronic exposure of these two populations of adult nonsmokers to a twofold difference in peak values of oxidant air pollution was not associated with significant differences in the described lung function tests or symptoms of chronic respiratory disease. A variety of explanations might account for these findings: (1) At current ambient values, oxidant air pollutants might not have had a harmful effect on the lungs of normal adults; however, the proved deleterious effects of relatively low concentrations of oxidant air pollutants on rubber, fabrics, and dyes, and certain plants, as well as the irritant effects on human eyes, places this conclusion in doubt. (2) These results might simply have reflected the known inadequacies of current techniques for the early detection of lung injury. (3) The significant health effects of oxidants might have been primarily important when the subjects were exposed during their growing years (34), and the subjects selected for this study probably had negligible exposure to oxidant air pollutants during their childhood. (4) Oxidant air pollution is a relatively recent insult to man's environment, and deleterious health effects of these pollutants might not yet be obvious. (5) Exposure to larger concentrations of oxidants might be demonstrably deleterious only when combined with the insult of smoking. (6) Chronic deleterious health effects might have been more closely related to the annual mean values of oxidants, which were essentially the same for both areas, than to the short-term peak values. (7) The population studied had quite a low prevalence of respiratory disease and might have represented a sample of the general population that is unusually resistant to the development of pulmonary disease, regardless of the causative agents. Persons with known or suspected predisposition to respiratory disease, such as α_1-antitrypsin-deficient homozygotes, might be more susceptible to pulmonary injury by oxidant air pollution.

Acknowledgment

The writers are grateful for the extensive assistance of Dr. John Finklea and Dr. Carl Shy as well as many other members of the Community Research Branch, Environmental Protection Agency, without whose help this study could not have been done.

References

1. Reid, D. D.: Air pollution as a cause of chronic bronchitis, Proc. Roy. Soc. Med., 1964, 57, 965.
2. Winkelstein, W., Jr., Kantor, S., Davis, E. W., Maneri, C. S., and Mosher, W. E.: The re-

43

lationship of air pollution and economic status to total mortality and selected respiratory system mortality in man, Arch. Environ. Health (Chicago), 1967, *14*, 162.

3. Holland, W. W., and Reid, D. D.: The urban factor in chronic bronchitis, Lancet, 1965, *1*, 445.

4. Mills, C. A.: Respiratory and cardiac deaths in Los Angeles smogs, Amer. J. Med. Sci., 1957, *233*, 379.

5. Breslow, L., and Goldsmith, J. R.: Health effects of air pollution, Amer. J. Public Health, 1958, *48*, 913.

6. Deane, M., Goldsmith, J. R., and Tuma, D.: Respiratory conditions in outside workers, Arch. Environ. Health (Chicago), 1965, *10*, 323.

7. Buell, P., Dunn, S. E., and Breslow, C.: Cancer of the lung and Los Angeles-type air pollution, Cancer, 1967, *201*, 2139.

8. Stern, A. C.: Air Pollution, ed. 2, vol. I, Academic Press, Inc., New York, 1968.

9. Young, W. A., Shaw, D. B., and Bates, D. V.: Effect of low concentrations of ozone on pulmonary function in man, J. Appl. Physiol., 1964, *19*, 765.

10. Bennett, G.: Ozone contamination of high altitude aircraft cabins, Aerospace Med., 1962, *33*, 969.

11. Goldsmith, J. R., and Nadel, J. A.: Experimental exposure of human subjects to ozone, J. Air Pollut. Contr. Assn., 1969, *19*, 3293.

12. Colley, J. R. T., and Holland, W. W.: Social and environmental factors in respiratory disease, Arch. Environ. Health (Chicago), 1967, *14*, 157.

13. Spicer, W. S.: Air pollution and meterologic factors, Arch. Environ. Health (Chicago), 1967, *14*, 185.

14. Bates, D. V., and Christie, R. U.: Respiratory function in disease, W. B. Saunders Company, Philadelphia, 1964, p. 374.

15. Damon, A.: Negro-white differences in pulmonary function, Hum. Biol., 1966, *38*, 380.

16. The Clean Air Quarterly, 1960–1968, State of California, Department of Public Health, Bureau of Air Sanitation.

17. Rose, G. A., and Blackburn, H.: Cardiovascular survey methods, W. H. O. Monogr. Ser., 1968, *56*, 1–188.

18. Hollingshead, A. B., Redlich, F. C.: Social class and mental illness, John Wiley & Sons, Inc., New York, 1958.

19. Air Quality in California, 1963–1967, California Air Resources Board, Sacramento, California.

20. Clinical Spirometry: Recommendations of the Section on Pulmonary Function Testing, Committee on Pulmonary Physiology, American College of Chest Physicians, Dis. Chest, 1963, *43*, 214.

21. Mills, J. N.: Variability of the vital capacity of the normal human subject, J. Physiol., 1949, *76*, 110.

22. Lemon, F. R., Walden, R. T., and Woods, R. W.: Cancer of the lung and mouth in Seventh-Day Adventists, Cancer, 1964, *17*, 486.

23. Lemon, F. R., and Walden, R. T.: Death from respiratory system disease among Seventh-Day Adventist men, J. A. M. A., 1966, *198*, 117.

24. Wynder, E. L., Lemon, F. R., and Mantel, N.: Epidemiology of persistent cough, Amer. Rev. Resp. Dis., 1965, *91*, 679.

25. Dysinger, P. W., and Lemon, F. R.: Pulmonary emphysema in nonsmoking population, Dis. Chest, 1963, *43*, 17.

26. Wynder, E. L., Lemon, F. R., and Bross, I. J.: Cancer and coronary artery disease among Seventh-Day Adventists, Cancer, 1959, *12*, 1016.

27. Brown, R., Woolcock, A. J., Vincent, N. S., and Macklem, P. T.: Physiological effects of experimental airway obstruction with beads, J. Appl. Physiol., 1969, *27*, 328.

28. Hogg, J. C., Macklem, P. T., and Thurlbeck, W. M.: Site and nature of airway obstruction in chronic obstructive lung disease, New Eng. J. Med., 1968, *278*, 1355.

29. DuBois, A. B., and Rogers, R. M.: Respiratory factors determining the tissue concentrations of inhaled toxic substances, Resp. Physiol., 1968, *5*, 34.

30. Buckley, R. D., and Loosli, C. G.: Effects of NO_2 on germ-free mouse lung, Arch. Environ. Health (Chicago), 1969, *18*, 588.

31. Darke, C. S., and Warrack, A. J. N.: Bronchiolitis from nitrous fumes, Thorax, 1958, *13*, 327.

32. Woolcock, A. J., Vincent, N. J., and Macklem, P. T.: Frequency dependence of com-

pliance as a test for obstruction in the small airways, J. Clin. Invest., 1969, *48*, 1097.

33. Mead, J.: The lung's "quiet zone," New Eng. J. Med., 1970, *282*, 1318.
34. Bartlett, D., Jr.: Postnatal growth of the mammalian lung: Influence of low and high oxygen tensions, Resp: Physiol., 1970, *9*, 58.

Long-Term Continuous Exposure to Sulfur Dioxide in Cynomolgus Monkeys

Yves Alarie, PhD; Charles E. Ulrich; William M. Busey, DVM, PhD; Alex A. Krumm; and Harold N. MacFarland, PhD

Cynomolgus monkeys were exposed to sulfur dioxide while a control group was exposed to filtered air. The exposure was for 24 hours a day, seven days a week, for 78 weeks. Frequent measurements were made to evaluate mechanical properties of the lung, distribution of pulmonary ventilation, diffusing capacity of the lung, and arterial blood oxygen tension. Hematological and clinical biochemical determinations were conducted. Microscopic evaluation of organs and tissues was conducted at termination of the exposure of the animals. No deleterious effect could be attributed to SO_2 exposure at concentrations of 0.14 to 1.28 ppm. Following 30 weeks of exposure to 4.69 ppm, an overexposure occurred in this group. This was followed by deterioration in pulmonary function which persisted during the following 48 weeks of observation, and alterations in pulmonary tissues were observed upon microscopic examination.

S ULFUR DIOXIDE, a common pollutant of urban air, has received wide attention in scientific literature; two reviews of the work done to delineate its effect on vegetation, animals, and man have been published.[1,2] From these reviews, it becomes obvious that the research undertaken had been mainly concerned with short-term exposures of animals[3-10] or man[11-18] and usually at higher concentrations than those encountered in urban atmospheres, with the exception of a study on guinea pigs exposed to 0.16 ppm.[19] The few studies conducted to elucidate long-term effects in animals[20-23] also involved high concentrations. More recently, however, comprehensive pulmonary function tests have been used in long-term exposure studies to determine the effects of sulfur oxides at low concentrations in dogs[24-25] and guinea pigs.[26]

In the present study, cynomolgus monkeys (*Macaca irus*) were exposed for a period of 18 months at concentrations ranging from 0.1 to 5.0 ppm. This range was selected because it includes those concentrations encountered in urban atmospheres and extends to the current threshold limit value of 5.0 ppm established by the American Conference of Governmental Industrial Hygienists.[27]

Materials and Methods

Animals and Exposure Conditions.—Five groups of animals were used in this study; each group consisted of nine animals composed of five males and four females or vice versa. One group was maintained as the control group and was exposed to filtered air, whereas the other four groups were exposed to SO_2 at the four approximate nominal concentrations of 0.1, 0.5, 1.0, and 5.0 ppm.

Young cynomolgus monkeys (*M irus*) weighing 1.6 to 3.0 kg (3.5 to 6.6 lb) were used in this study. They were individually caged and held in a quarantine area. The animals were tested twice for tuberculosis during an eight-week period and then transported from the quarantine area to laboratory-animal holding rooms where they were observed for a period of two weeks. During the following 8 to 12 weeks, the animals were retested for tuberculosis, had a chest x-ray film taken, and passed through a screening procedure which consisted of evaluation of pulmonary function, electrocardiographic findings, and hematologic and clinical chemistry studies. Animals having values

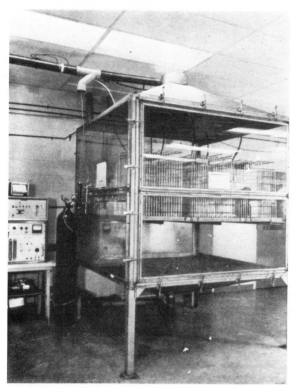

Fig 1.—Exposure chamber housing nine cynomolgus monkeys. Delivery and monitoring apparatus for SO_2 is also shown.

variations present from week to week. Then the exposure to SO_2 was initiated and continued for a period of 78 weeks.

Each day the exposure was interrupted twice for a period of ten minutes for feeding the animals and cleaning the chambers. Exposures were also interrupted for the duration of the particular test on days when physiological measurements were made.

Samples of each chamber's atmosphere were taken once daily with an impinger and analyzed according to the method of West and Gaeke.[28] Also SO_2 analyzers (Melpar SR-1100-1) were used for semicontinuous analysis of chamber concentration of SO_2 and calibrated against the wet chemical method.[28] One analyzer was shared between the 0.1- and 0.5-ppm SO_2 exposure chambers and another was shared between the 1.0- and 5.0-ppm SO_2 chambers.

Measurements Made on Animals.—Body weights were recorded weekly during the eight-week preexposure period and the 78 weeks of exposure. Observation for survival and gross signs of behavioral abnormality were made daily. Pulmonary function tests included the measurements listed below. All were performed without anesthesia after 10 to 20 minutes of adaptation with the animal in the restraining chair in a sitting position with the face mask attachment.

Mechanical Properties of the Lung.—Measurements of total respiratory system flow resistance during inspiration (Rrs[i]) and during expiration (Rrs[e]) were made every week during the preexposure control period, every week during the first nine weeks of exposure, and every four weeks thereafter, employing a previously published method.[29] Using previously published techniques,[30] measurements of tidal volume (V_T), respiratory rate (RR), minute volume (MV), dynamic compliance of the lung (Cdyn[1]), pulmonary flow resistance (R1), and work of breathing during inspiration (W[i]) and during expiration (W[e]) were made five times during the pre-

within normal limits were accepted. During this period, animals were also placed in a restraining chair with a face mask to adapt them to the pulmonary function testing procedures.

Animals accepted for studies were then transferred to individual exposure cages and placed in stainless steel and glass chambers. One of the exposure chambers used in this experiment is shown in Fig 1; the volume of each is approximately 6.5 cu m.

The air supplied to each exposure chamber passed through a charcoal bed and an absolute filter. The temperature was maintained at 72 ± 2 F and the relative humidity at 50% ± 5%. The airflow through each chamber was slightly in excess of 1,000 liter/min.

After placing the animals in the exposure chambers, a period of two weeks was allowed for adaptation. During the following eight weeks, all groups were exposed to filtered air only and physiological measurements were made to establish baseline values and normal

48

exposure control period, every two weeks during the first ten weeks of the exposure period, and every four weeks thereafter.

Distribution of Ventilation.—The distribution of pulmonary ventilation was evaluated by the nitrogen washout technique as previously described.[31] These measurements were made on six occasions during the preexposure control period, every two weeks during the first four weeks of the exposure period, and every four weeks thereafter.

Diffusing Capacity of the Lung.—The carbon monoxide uptake and the diffusing capacity of the lung were measured by a rebreathing technique based on the same principles as those described by Bates,[32] Lewis et al,[33] Kruhoffer,[34] and Long and Hatch[35] after modification for use in monkeys. In doing so, the face mask and chair, which have been previously described,[29,31] were used. These measurements were made according to the same schedule as the distribution of ventilation measurements.

Arterial Blood Gas.—Arterial blood was obtained from the femoral artery for measurements of oxygen tension (Pao_2), carbon dioxide tension ($Paco_2$), and pH. Measurements were made at 37.5 C utilizing oxygen (Po_2) and carbon dioxide (Pco_2) tension and pH electrodes on a radiometer, as previously described.[36] Each sample of blood was obtained 10 to 20 minutes after injection of phencyclidine hydrochloride for tranquilization of the animal. Measurements were made once during the preexposure period, every eight weeks during the first 24 weeks of the exposure period, and every 12 weeks thereafter.

Hematologic and Clinical Chemistry Studies.—Blood obtained from the femoral artery was used for the determination of the following values: hematocrit, hemoglobin, total erythrocyte and leukocyte counts, lymphocytes, segmented neutrophils, blood urea nitrogen (BUN), total bilirubin, serum total protein, serum albumin, serum electrolytes, serum glutamic oxaloacetic transaminase (SGOT), serum glutamic pyruvic transaminase (SGPT), serum lactic acid dehydrogenase, and serum alkaline phosphatase. We made the determinations, once during the preexposure control period and at weeks 11, 16, 25, 49, and 77 of the exposure period.

Statistical Analysis.—All data obtained were stored on digital computer system (IBM 1800) and statistical analyses were performed by this system. For each preexposure interval, data were analyzed to delineate baseline values, frequency distribution characteristics, mean values, standard deviation, and coefficient of varia-

tion of each individual group, of the five groups combined, and of differences between males and females. Regression analyses were used on the measured parameters when sufficient data points (greater than eight) were available. The proper polynomial regression function was chosen on the basis of the best fit for the control or one of the treatment groups. In these analyses, individual data points were used for each time interval. Comparisons of the regression coefficients between groups were made by Student's t-test, and confidence intervals were calculated as previously described.[30,31] In some instances a cumulative sum technique (CUSUM)[37-39] was used, via extension, to establish the time of occurrence of significant differences between the preexposure control period and the exposure period for a particular group or between the control group and any of the exposed groups. In performing statistical comparisons, $P < .05$ was chosen in all cases for accepting or rejecting significant differences between groups.

Killing Procedures.—Following the 78 weeks of exposure, the nine animals were killed with an overdose of pentobarbital sodium followed by exsanguination. Complete necropsies were performed on each animal, and specimens of the following tissues were fixed in 4% neutral buffered formaldehyde solution: trachea, peribronchial lymph node, heart, liver, and kidney. After fixation, 6μ hematoxylin-eosin-stained slides were prepared. In an inflated state, the lungs were removed from the animals and were perfused intratracheally with a volume of neutral buffered formaldehyde solution equal to the V_T of the animal. Slides were prepared from each of the seven lobes of the lungs.

Results

Sulfur Dioxide Concentrations.—The chamber concentrations of SO_2 as determined are presented in the Table. The arithmetic means of the monthly averages for each of the chambers and the standard deviations were 0.14 ± 0.05, 0.64 ± 0.25, 1.28 ± 0.42, and 4.69 ± 0.34 ppm. An accidental overexposure to SO_2 occurred after 30 weeks of exposure in the chamber in which animals were exposed to 4.69 ppm. The concentration during the overexposure was estimated to be not higher than 1,000 ppm and not lower than 200 ppm. The animals remained in that atmosphere for a period of one hour. Following this accident, the animals were not reexposed to SO_2 but were kept in the chamber

49

with filtered air for the following 48 weeks and physiological parameters were measured as originally scheduled until termination of exposure.

Measurements of Biological Parameters: Body Weight and Survival.—Analysis of the growth curve for each group revealed no significant differences between the control group and the exposed groups during the exposure to SO_2, although the group exposed to 0.14 ppm of SO_2 had a slightly higher body weight than the control group at the beginning of the experiment. The body weight growth curves for the control group and the group exposed to 4.69 ppm of SO_2 are presented in Fig 2. One death occurred in the control group and one in the group exposed to 0.64 ppm of SO_2. The death in the group exposed to 0.64 ppm of SO_2 was caused by intestinal infections accompanied by diarrhea and dehydration, whereas the death occurring in the control group was caused by gastric distention. No deaths occurred in the group exposed to 4.69 ppm of SO_2, which was accidentally overexposed, or in the groups exposed to 0.14 or 1.28 ppm of SO_2.

Mechanical Properties of the Lungs.—The V_T values obtained during the preexposure period and the 78-week exposure period for the control group and the groups exposed to SO_2 were compared. The groups exposed to 0.14, 1.28, and 4.69 ppm of SO_2 had increases in V_T comparable to the increase observed in the control group, but the group exposed to 0.64 ppm of SO_2 had a slight but significantly lower rate of increase in V_T than the control group. The RR values for the groups exposed to 0.14, 0.64, 1.28, and 4.69 ppm were comparable to the RR value of the control group. Comparable fluctuations were observed in RR in all groups, as were previously observed. The Rrs(i) and Rrs(e) values obtained during the preexposure period and the 78 weeks of exposure for the control group and the group exposed to 4.69 ppm of SO_2 are presented in Fig 3 and 4, respectively. There was a general trend toward decreasing values of both Rrs(i) and Rrs(e) in all groups, and no significantly higher values in Rrs(i) or Rrs(e) were observed in the groups exposed to SO_2 than in the control group. For the group exposed to 4.69 ppm of SO_2 there was, however, more variation in Rrs(i) values following the overexposure accident than in the Rrs(i) values of the control group. Values for R1 followed the same trend, and no significant differences were observed between the control group and the groups exposed to SO_2. At the beginning of the exposure, Cdyn(1) values were comparable in all groups and were as follows: 7.86 ml/cm H_2O for the control group 8.00 mg/cm H_2O, 7.91 mg/cm H_2O, and 7.02 mg/cm H_2O, and 7.15 mg/cm H_2O for the groups exposed to 0.14, 0.64, 1.28, and 4.69 ppm of SO_2, respectively. The values for Cdyn(1) remained within a narrow range during the entire exposure period, and no significant trends were noted with exposure to SO_2. Similarly, values for W(i)/ ml V_T and W(e)/ml V_T exhibited no significant changes due to SO_2 exposure.

Distribution of Ventilation, Diffusion, and Arterial Blood.—Values for the number of breaths between the beginning and end of the test it took to reach an end-expiratory N_2 concentration of 1% ($N[1\% \ N_2]$) for nitrogen washout measured during the preexposure period and the 78-week exposure period are presented in Fig 5. The data were analyzed and a third-degree polynomial curve fitting is presented for the control group and the group exposed to 4.69 ppm of SO_2. There were no significant differences due to SO_2 in the groups exposed to 0.14, 0.64, and 1.28 ppm of SO_2. For the group exposed to 4.69 ppm of SO_2, a definite increase in $N(1\% \ N_2)$ values occurred following the overexposure accident. To further delineate this increase in $N(1\% \ N_2)$ values, the data were analyzed with the CUSUM technique[37-39] and the results are presented in Fig 6. In Fig 6, *top*, the CUSUM of the control group indicated that no change took place during the preexposure or exposure period, whereas the CUSUM of the group exposed to 4.69 ppm of SO_2 indicated a significant increase in $N(1\% \ N_2)$ values starting around week 33. Figure 6, *bottom*, also shows that a significant change occurred around week 33. Also measured during the nitrogen washout measurements were the time from the beginning of the test to reach an end-expiratory N_2 concentration of 1% ($t[1\% \ N_2]$) and the cumulative V_T from the beginning to the end of the test ($CV_T[1\% \ N_2]$), and in both instances anal-

50

		Analytical Concentrations of SO$_2$ in Exposure Chambers			
Date	No.	First Level of Exposure, $\overline{X} \pm$ (SD)	Second Level of Exposure, $\overline{X} \pm$ (SD)	Third Level of Exposure, $\overline{X} \pm$ (SD)	Fourth Level of Exposure, $\overline{X} \pm$ (SD)
7/17/67-7/31/67	15	...	...	0.97 ± (0.25)	4.68 ± (0.35)
7/26/67-7/31/67	6	...	0.49 ± (0.06)	...	...
8/1/67-8/31/67	31	...	0.50 ± (0.03)	1.03 ± (0.07)	4.89 ± (0.24)
8/8/67-8/31/67	24	0.12 ± (0.05)	...	...	...
9/1/67-9/30/67	30	0.10 ± (0.01)	0.50 ± (0.03)	1.02 ± (0.13)	5.03 ± (0.19)
10/1/67-10/31/67	31	0.12 ± (0.03)	0.51 ± (0.13)	1.04 ± (0.22)	5.28 ± (1.62)
11/1/67-11/30/67	30	0.10 ± (0.01)	0.51 ± (0.04)	1.06 ± (0.14)	4.45 ± (0.53)
12/1/67-12/31/67	31	0.11 ± (0.02)	0.45 ± (0.07)	1.13 ± (0.24)	4.32 ± (0.52)
1/1/68-1/31/68	31	0.16 ± (0.26)	0.54 ± (0.23)	1.18 ± (0.10)	4.53 ± (0.61)
2/1/68-2/11/68	11	...	...	...	4.38*± (0.17)
2/1/68-2/29/68	29	0.11 ± (0.02)	0.49 ± (0.08)	1.09 ± (0.22)	...
3/1/68-3/31/68	31	0.12 ± (0.05)	0.51 ± (0.03)	0.94 ± (0.13)	...
4/1/68-4/30/68	30	0.10 ± (0.02)	0.47 ± (0.06)	1.01 ± (0.13)	...
5/1/68-5/31/68	31	0.14 ± (0.03)	0.81 ± (0.04)	1.72 ± (0.16)	...
6/1/68-6/30/68	30	0.13 ± (0.03)	0.70 ± (0.03)	1.51 ± (0.20)	...
7/1/68-7/31/68	31	0.18 ± (0.05)	0.75 ± (0.05)	1.53 ± (0.11)	...
8/1/68-8/31/68	31	0.24 ± (0.03)	1.22 ± (0.06)	2.21 ± (0.11)	...
9/1/68-9/30/68	30	0.25 ± (0.02)	1.25 ± (0.04)	2.06 ± (0.14)	...
10/1/68-10/31/68	31	0.23 ± (0.02)	0.99 ± (0.04)	2.00 ± (0.09)	...
11/1/68-11/30/68	30	0.11 ± (0.01)	0.49 ± (0.05)	0.93 ± (0.09)	...
12/1/68-12/31/68	31	0.11 ± (0.03)	0.48 ± (0.05)	0.98 ± (0.10)	...
1/1/69-1/10/69	10	...	...	0.98 ± (0.09)	...
1/1/69-1/17/69	17	...	0.46 ± (0.05)	...	...
1/1/69-1/30/69	30	0.11 ± (0.03)	...	...	...
$\overline{\overline{X}} \pm$ (SD)		0.14 ± (0.05)	0.64 ± (0.25)	1.28 ± (0.42)	4.69 ± (0.34)

* Overexposure accident occurred on Feb 11, 1968, after 30 weeks of exposure. In this incident, the concentration was estimated to be as high as 1,000 ppm and not lower than 200 ppm for a period of one hour. Following this accident, the animals remained in the chamber for the following 48 weeks until Jan 9, 1969, when they were killed; they were not further exposed to SO$_2$ during that period.

ysis of the data similar to $N(1\% \ N_2)$ revealed a definite increase in those values in the group exposed to 4.69 ppm of SO$_2$ after the overexposure accident. All groups had comparable diffusing capacities of the lung for carbon monoxide prior to exposure and no significant trends were observed during the exposure to SO$_2$. There was a significant decrease in Pao$_2$ values in the group exposed to the 4.69 ppm of SO$_2$ during the exposure period. A small decrease occurred in the group exposed to 1.28 ppm of SO$_2$ at the beginning of exposure, but in this group the values for Pao$_2$ were comparable to the control group during the remainder of the exposure period. No significant changes occurred in the groups exposed to 0.14 or 0.64 ppm of SO$_2$. No changes were observed in Paco$_2$ values in the groups exposed to SO$_2$, and arterial pH values remained essentially

constant with the exception of a low pH value of 7.32 observed at week 77 in the group exposed to 4.69 ppm of SO$_2$.

Hematological and Clinical Biochemical Determinations.—The values obtained for the hematocrit reading, hemoglobin, and erythrocyte count were within normal limits in all groups during the exposure to SO$_2$. High mean values for leukocytes of 11,000 to 15,000 were observed in all groups during the preexposure period. These values decreased during the exposure and reached 8,700 to 9,700 at week 77. A high value of 13,900 was observed at week 34 for the group exposed to 4.69 ppm of SO$_2$. This was three weeks after the overexposure accident which occurred in this group's chamber. Values for lymphocytes and neutrophils were within normal limits in all groups. Blood urea nitrogen and total bilirubin determina-

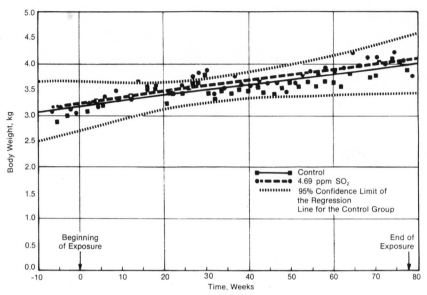

Fig 2.—Body weight data prior to and during exposure to SO_2. Each **point** represents mean of nine animals, and curves were fitted with first-degree polynomial. Regression lines for groups exposed to 0.14, 0.64, and 1.28 ppm of SO_2 are not presented as they fell between 95% confidence limits of regression line for control group.

Fig 3.—Rrs(i) data prior to and during exposure to SO_2. Each **point** represents mean of nine animals, and curves were fitted with first-degree polynomial. Curves for groups exposed to 0.14, 0.64, and 1.28 ppm of SO_2 are not presented since they lay between 95% confidence limits of regression line for control group.

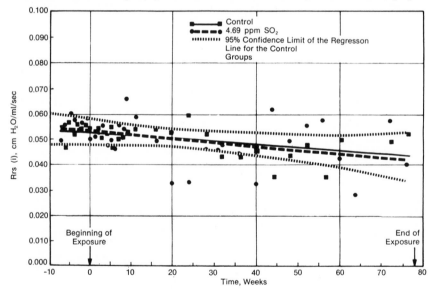

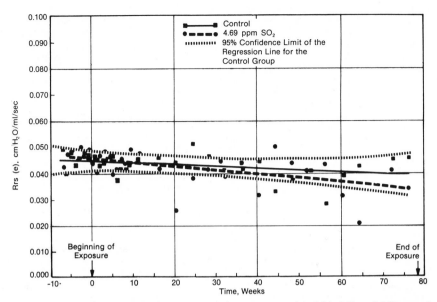

Fig 4.—Rrs(e) data prior to and during exposure to SO₂. Each **point** represents mean of nine animals, and curves were fitted with first-degree polynomial. Curves for groups exposed to 0.14, 0.64, and 1.28 ppm of SO₂ are not presented since they lay between 95% confidence limits of regression line for control group.

Fig 5.—N(1% N₂) data prior to and during exposure to SO₂. Each **point** represents mean of nine animals and curves were fitted with third-degree polynomial. Regression lines for groups exposed to 0.14, 0.64, and 1.28 ppm of SO₂ are not presented as they were comparable to control group.

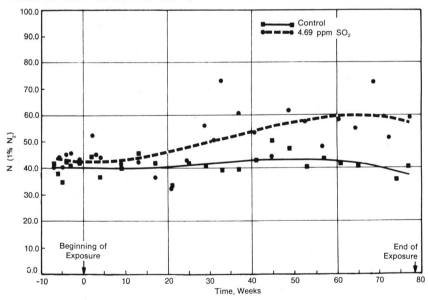

tions did not indicate any particular trend and a considerable variation was present. Serum total protein and albumin values were comparable in all groups and remained essentially constant during the exposure period. Similarly, serum electrolytes, including $Na+$, $K+$, $Cl-$, and $Ca++$, were comparable in all groups and remained within narrow limits during the exposure period. Serum glutamic oxaloacetic transaminase determinations remained within normal limits and considerable variation was present. For SGPT, no value higher than that of a control was observed but all groups exhibited low values at various time intervals. The results from measurements of serum lactic acid dehydrogenase and serum alkaline phosphatase indicated that all groups were comparable at the beginning of the exposure, and a definite trend toward lower values was present for all groups as the exposure progressed.

Terminal Body Weight, Organ Weight, and Organ-Body Weight Ratios.—Analysis of variance of terminal body weight, organ weight, and organ-body weight ratios obtained separately for males and females did not reveal any significant differences between the control and exposed groups.

Histopathological Examination.—*Control Group.*—Microscopic examinations of the lungs from the nine monkeys in the control chamber did not reveal any meaningful alterations. A slight to moderate degree of perivascular and peribronchiolar pigmentation was present in some of the animals in this group. This pigment was crystalline in form and generally brownish to black in color. The majority of this pigment had been phagocytized by tissue macrophages. A similar pigment was present in the peribronchial lymph nodes of these animals. The lungs from these animals were within the limits of normal histological variation and no evidence of occult disease was present.

Three instances of submucosal histiocytic infiltration and submucosal lymphoid hyperplasia were seen in the trachea. Here again, these microscopic alterations were slight in severity and are frequently seen in this species of primate. The livers were within the limits of normal histological variations for this species of monkeys. A few instances of slight hepatocyte pigmentation and Kupf-

fer's cell pigmentation were seen. The kidneys from these animals were within the limits of histological variation and only a few incidental microscopic alterations were present. A minimal to moderate degree of interstitial nephritis was present in two of the males in this group.

One animal from this group died of gastric distention during the 77th week of exposure.

Group Exposed to 0.14 ppm of Sulfur Dioxide.—The lungs from all the animals in this group were comparable to those from the controls. Peribronchial and perivascular pigmentation, similar to that seen in the controls, was present from a slight to moderate degree in all nine animals. A parasitic granuloma involving some alveoli was present in one animal.

Crystalline pigment was present from a minimal to moderately severe degree in the peribronchial lymph nodes in six of the nine animals in this group. No meaningful alterations were seen in the tracheae, livers, or hearts.

Group Exposed to 0.64 ppm of Sulfur Dioxide.—The lungs from all the surviving animals within this group were within the limits of histological variation and compared well with the controls. In one animal, which died from gastroenteritis, a moderate pulmonary edema was present. The pulmonary edema seen in this animal was probably terminal and resulted from protracted clinical gastroenteritis.

With the exception of a moderate degree of submucosal lymphoid hyperplasia in one animal, the tracheae from the animals in this group were not remarkable. Minimal to moderate amounts of crystalline pigment were seen in the peribronchial lymph nodes in eight of nine animals in this group. The hearts and livers from the animals in this group were comparable to the controls.

Group Exposed to 1.28 ppm of Sulfur Dioxide.—Peribronchial and perivascular pigment, similar to that seen in the controls, was present in the lungs from the majority of the animals in this group. No other meaningful alterations were present in the lungs.

A minimal to slight degree of submucosal lymphoid hyperplasia was present in the tracheae of two animals in this group. Crystalline pigment was present in the peribron-

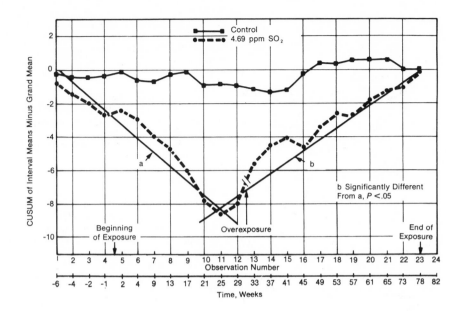

Fig 6.—**Top**, CUSUM transformation of data presented in Fig 5 for control group and group exposed to 4.69 ppm of SO₂. Data for control group remained essentially constant. Two distinct segments are present for group exposed to 4.69 ppm indicating a difference between data prior to and following overexposure. **Bottom**, CUSUM transformation of data presented in Fig 5 based on time-paired differences between control group and group exposed to 4.69 ppm of SO₂. Two distinct segments are present indicating that control group and exposed group were comparable before overexposure but significantly different following overexposure.

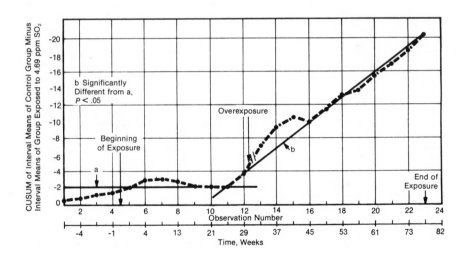

55

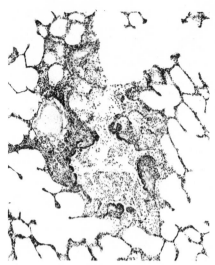

Fig 7.—Lung of animal overexposed to SO₂ showing alveolitis and bronchiolitis with pneumocyte hyperplasia (hematoxylin-eosin reduced from ×35).

chial lymph nodes in five of nine animals in this group. Two animals were affected with a minimal focal interstitial nephritis. The hearts and livers from all of the nine animals in this group did not show any meaningful alterations and were comparable to the controls.

Group Exposed to 4.69 ppm of Sulfur Dioxide and Overexposure Accident.—Histopathological alterations were seen in the lungs from all of these animals. Microscopic alterations were present in the respiratory bronchioles, alveolar ducts, and alveolar sacs. Focally distributed throughout all lobes of the lungs, several groups of alveoli contained moderate amounts of proteinaceous material and numerous alveolar macrophages. These alveoli were generally those opening directly off of the respiratory bronchioles. The alveolar walls were moderately thickened and infiltrated with histiocytes. Multinucleated giant cells were present in some alveolar lumina. The walls of the respiratory bronchioles associated with these foci were moderately infiltrated with histiocytes and leukocytes. Moderate hyperplasia of the bronchiolar epithelium was also present. Frequently the lumina of the respiratory bronchioles were plugged with proteinaceous

material, macrophages, and leukocytes.

In eight of the nine animals, bronchiectasis and bronchiolectasis were present to a moderate degree. The walls of many of the affected bronchioles were moderately thickened.

Even though the overexposure had occurred 48 weeks before the termination of exposure of these animals, there was little evidence of fibrosis. These pulmonary lesions appeared active and would be classified as subacute. The alterations described above are demonstrated by Fig 7.

Microscopic alterations apparently related to the exposure to SO₂ were also seen in the liver. These changes were characterized by diffuse and focal hepatocyte vacuolation varying in severity from moderate to severe. This hepatocyte vacuolation was seen in all nine animals in this group and was not seen in any of the control animals.

Moderate amounts of crystalline pigment, similar to that seen in controls, were present in the peribronchial lymph nodes. In the kidneys, a focal, minimal to slight interstitial nephritis was seen in two animals. The tracheae and hearts from all the animals in the group were within the limits of histological variation.

Comment

We reported previously[26] that no detrimental changes to the pulmonary system in guinea pigs could be detected after one year of exposure to concentrations ranging from 0.13 to 5.72 ppm of SO₂. The results presented in this report indicate that no detrimental changes in monkeys can be attributable to continuous exposure to SO₂ at 0.14, 0.64, or 1.28 ppm for a period of 78 weeks. The group exposed to 4.69 ppm for a period of 30 weeks did not seem to be adversely affected. The overexposure accident was followed by a definite deterioration in pulmonary function which was demonstrated by the changes in the distribution of pulmonary ventilation as measured by the nitrogen washout technique. The microscopic findings in this group indicated that not only the lungs had been adversely affected, but also that alterations in the liver, similar to those seen in guinea pigs exposed to 5.72 ppm,[26] were present. Another group of animals is now being exposed to 5.0 ppm for 78 weeks.

There are only a few reports involving chronic exposure of normal animals to sulfur oxides in which pulmonary function tests were included to delineate the effects on the pulmonary system. In a study by Goldring et al,[26] hamsters exposed chronically but not continuously to a very high concentration of SO_2 had only mild bronchitic lesions and relatively minor changes in the mechanical properties of the lung. These authors concluded that even at high concentrations, SO_2 failed to produce an important effect on the experimental production of obstructive lung disease in this species. Lewis et al[25] compared the effects of 5.1 ppm of SO_2, 5.1 ppm of SO_2 plus $835\mu g/cu$ m of H_2SO_4, or $0.755\mu g/cu$ m of H_2SO_4 in dogs impaired by prior exposure to 26 ppm of nitrogen dioxide with dogs not impaired. They did not report a statistically significant change in pulmonary function between the impaired and unimpaired control groups and any one of the exposed groups that received SO_2, H_2SO_4, or a mixture of both. They reported some differences between the various groups and suggested that the oxides of sulfur appeared to affect the unimpaired dogs more adversely than the impaired animals.[25] A study by Vaughan et al[24] involved dogs exposed 16 hours a day for a period of 18 months to a combination of 0.5 ppm of SO_2 and $100\mu g/cu$ m of H_2SO_4 mist. These authors concluded that after 18 months of exposure there were no detectable changes in the mechanical properties of the lung or in the diffusing capacity of the lung in the exposed animals. These results are similar to those obtained previously in guinea pigs[26] and those presented in this article. Vaughan et al[24] cautioned that their failure to detect impairment of pulmonary function might have been due to several factors, including (a) actual lack of effect due to exposure, (b) inability of the methods to detect subtle abnormalities, (c) quickly reversible effects due to removal of the animals from the contaminated atmosphere for measurements of pulmonary function, and (d) a large range of normal values in the control population necessitating large differences to validate the changes statistically. Also, in the studies by Vaughan et al[24] and Lewis et al,[25] measurements of pulmonary function were made during or following anesthesia, which in itself may influence the results greatly. Thus, each of these points mentioned above must be evaluated before it can be concluded with any degree of certainty that the first, that is, actual lack of effects from exposure to SO_2, is the most probable choice. In such an attempt, the ability of the pulmonary function tests used in this study to detect subtle abnormalities must be considered first. Measurements of the mechanical properties of the lung, including V_T, RR, respiratory system flow resistance, R1, Cdyn(1), W(i), and W(e), have been made by numerous investigators in order to evaluate the effects of acute or subacute exposure to various pulmonary irritants, bronchoconstrictors,[3,40-43] and inert dusts[44,45] or to evaluate patients with obstructive respiratory diseases.[46-50] There is no doubt that these parameters are sensitive indices of respiratory impairment, particularly in acute exposure in which direct or reflex bronchoconstriction may occur[3]; but, when subtle changes have occurred, principally in the peripheral airways, parameters such as airway resistance can fail to indicate obstruction, as recently reported by Hogg et al.[50] As recent literature attests, there is also an obvious lack of knowledge about the correlation between subtle microscopic alterations in the lung and concomitant changes in these physiological parameters. In a review of the various techniques to assess the distribution of ventilation or distribution of inspired air in the lung, Bouhuys[51] has pointed out the usefulness of the nitrogen washout technique in detecting subtle changes due to small airway obstruction which would have otherwise gone undetected by measurements of airway resistance.[50] Bouhuys states that this measurement has proven a valuable tool in assessing the pathogenesis of bronchial asthma, emphysema, and worker's byssinosis.[51-54] Significant distribution changes could be demonstrated with inhaled histamine when subjective symptoms in humans were absent.[55] The technique has also been used in dogs exposed to phosgene and has revealed deterioration in pulmonary function.[43] This technique, adapted for the cynomolgus monkey, has been successful in detecting the effect of the overexposure of SO_2 sustained in the group exposed to 4.69 ppm.

The measurement of the diffusing capacity of the lung is probably the most difficult to evaluate. Forster[56-57] indicated the various factors governing the lung's diffusing capacity for carbon monoxide, the discrepancies between the methods used for evaluating this parameter, and the factors to be considered in the interpretation of the results. On the other hand, Long and Hatch[35] and Rinehart and Hatch[58] have used CO uptake measurements and have concluded that, following exposure to low levels of pulmonary irritants, a decrease could be observed without observable microscopic alterations in the lung. In addition, exposure to ozone has been reported to result in significant decreases in the diffusing capacity of the lung,[59,60] and reduction in CO uptake has been reported in cases of emphysema.[32] In a previously published article,[26] we reported a higher diffusing index in the group of guinea pigs exposed to 5.72 ppm of SO_2. This finding correlated well with the microscopic examination of the lung which revealed a much lower incidence of spontaneous pulmonary disease at this exposure level than in the control group.

While it is true that transient effects could not be detected by pulmonary function tests performed on animals removed from the exposure atmosphere, our study was directed toward the elucidation of effects resulting from chronic exposure. Thus, our experimental design purposely precluded the detection of immediate and possibly reversible effects of SO_2 reflex bronchoconstriction.[7,10] Moreover, on the basis of published data from short-term human exposures,[11-13,17] one would not expect consistent changes in the mechanical properties of the lung at concentrations up to 5 ppm. In guinea pigs, probably the most sensitive species studied in acute exposure to SO_2, an increase of only 10% in R1 is observed at 0.16 ppm of SO_2[19] and 25% at 26 ppm,[3] with a rapid return toward control values upon termination of the exposure.

Lastly, Vaughan et al[24] and Lewis et al[25] reported large variations, and large changes were needed to detect a statistically significant change. In our study, the five groups of animals had mean values which were very close for all parameters before the beginning of exposure. In addition, a sufficiently long preexposure observation period assured the establishment of baseline control values in each group, and measurements were made frequently during the entire 78 weeks of exposure to insure a higher chance of detecting any change.

In view of the above, we believe that when we have stated that there were no deleterious changes in the pulmonary system, there was indeed a genuine lack of effect ascribable to SO_2. This is also substantiated by the fact that histopathological examination of the lungs confirmed our observations with pulmonary function tests.

Finally, we must point out that negative results obtained with SO_2 on R1 and Rrs in monkeys exposed for a period of 18 months to SO_2 in no way contradict the findings in guinea pigs by Amdur and Mead[3] and Amdur[19] and in man by several authors[11-13,17] who found an increase, although relatively small, in R1 with short-term exposure up to 5 ppm SO_2. Indeed, in these studies, a slight bronchoconstriction was apparent during exposure.[7] In studies on long-term chronic exposure, increases in R1 or Rrs are more likely to be caused by tissue deterioration than by sustained reflex bronchoconstriction, and it is important to distinguish the two phenomena. Furthermore, the use of R1 or Rrs may not be quite appropriate to detect deterioration in pulmonary function if the effects are at the periphery, since a large increase, up to 100%, in the peripheral resistance would contribute only a small fractional increase in R1 or Rrs.[50] As pointed out by Hogg et al,[50] measurement of distribution of ventilation would be more appropriate to detect pulmonary damage in these instances. This was also found by Rossing[43] when exposing dogs to phosgene for periods of 13 weeks and also reported here in cynomolgus monkeys exposed to 4.69 ppm of SO_2, following the overexposure accident, and in a previous report[31] in cynomolgus monkeys exposed to NO_2.

The present findings in cynomolgus monkeys and the results reported earlier for the guinea pig[26] indicate that SO_2 in concentrations up to 1 ppm does not cause alterations in pulmonary function, nor are detrimental changes seen on microscopic examination of pulmonary tissues. In this connection, it should be remembered that Vaughan et al[24]

58

also obtained normal results in dogs chronically exposed to low concentrations of SO_2; these authors have not presented histopathological findings as yet. Thus, the results of studies in a rodent, the dog, and a primate have all been normal.

In urban atmospheric pollution, populations are exposed to SO_2 in combination with other gaseous and particulate pollutants, and the possibility of synergistic effects under these circumstances must be considered. In our studies, exposure of two groups of cynomolgus monkeys to fly ash at 0.16 and 0.46 mg per cubic meter for a period of 18 months did not produce significant changes in pulmonary function in these animals[61]; other groups are now undergoing exposure to combinations of SO_2 and fly ash. It should be possible, therefore, to elucidate the possible synergistic action of these two pollutants in studies on chronic exposure. In short-term studies, no synergistic action was reported between SO_2 and fly ash, while certain metallic salts when added to SO_2 were shown to increase the response of guinea pigs.[62,63]

Dr. Alarie is under special fellowship 1-FO3-ESH6198-01 from the National Institute of Environmental Health Sciences.

The data reported in this article have resulted from the Electric Research Council's Air Pollution Research Program financed by the Edison Electric Institute with assistance from the Tennessee Valley Authority, the American Public Power Association, and Bituminous Coal Research, Inc.

Nonproprietary and Trade Names of Drug

Phencyclidine hydrochloride—*Sernylan*.

References

1. Negherbon WO: *Sulfur Dioxide, Sulfur Trioxide, Sulfuric Acid and Fly Ash: Their Nature and Their Role in Air Pollution*, research project RP-62, EEI publication 66-900. Falls Church, Va, Hazleton Laboratories Inc, 1966.

2. *Air Quality Criteria for Sulfur Oxides.* US Dept of Health, Education, and Welfare, 1968.

3. Amdur MO, Mead J: A method for studying the mechanical properties of the lungs of unanesthetized animals: Application to the study of respiratory irritants, in *Proceedings of the Third National Air Pollution Symposium.* Los Angeles, 1955, pp 150-159.

4. Balchum OJ, Dybicki J, Meneely GR: Pulmonary resistance and compliance with concurrent radioactive sulfur distribution in dogs breathing $S^{35}O_2$. *J Appl Physiol* 15:62-66, 1960.

5. Frank NR, Speizer FE: SO_2 effects on the respiratory system in dogs. *Arch Environ Health* 11:624-634, 1965.

6. Amdur MO: The physiological response of guinea pigs to atmospheric pollutants. *Int J Air Pollut* 1:170-183, 1959.

7. Nadel JA, Salem H, Pamplin B, et al: Mechanisms on bronchoconstriction during inhalation of sulfur dioxide. *J Appl Physiol* 20:164-167, 1965.

8. Amdur MO: The influence of aerosols upon the respiratory response of guinea pigs to sulfur dioxide. *Industr Hyg Assoc Quart* 18:149, 1957.

9. Amdur MO, Underhill D: The effects of various aerosols on the response of guinea pigs to sulfur dioxides. *Arch Environ Health* 16:460-468, 1968.

10. Widdicombe JG: Respiratory reflexes from the trachea and bronchi of the cat. *J Appl Physiol* 123:55-70, 1954.

11. Frank NR: Studies on the effects of acute exposure to sulphur dioxide in human subjects. *Proc Roy Soc Med* 57:1029-1033, 1964.

12. Frank NR, Amdur MO, Whittenberger JL: A comparison of the acute effects of SO_2 administered alone or in combination with NaCl particles on the respiratory mechanisms of healthy adults. *Int J Air Pollut* 8:125-133, 1964.

13. Frank NR, Amdur MO, Worcester J, et al: Effects of acute controlled exposure to SO_2 on respiratory mechanics in healthy male adults. *J Appl Physiol* 17:252-258, 1962.

14. Lawther PJ: Effects of inhalation of SO_2 on respiration and pulse rate in normal subjects. *Lancet* 2:745-748, 1955.

15. Toyoma T, Nakamura K: Synergistic response of hydrogen peroxide aerosols and sulfur dioxide to pulmonary airways resistance. *Industr Health* 2:34-45, 1964.

16. Speizer FE, Frank NR: A comparison of changes in pulmonary flow resistance in healthy volunteers acutely exposed to SO_2 by mouth and by nose. *Brit J Industr Med* 23:75-79, 1966.

17. Burton GG, Corn M, Gee JBL, et al: Response of healthy men to inhaled low concentrations of gas-aerosol mixtures. *Arch Environ Health* 18:681-692, 1969.

18. Snell RE, Luchsinger PC: Effects of sulfur dioxide on expiratory flow rates and total respiratory resistance in normal human subjects. *Arch Environ Health* 18:693-698, 1969.

19. Amdur MO: Respiratory absorption data and SO_2 dose-response curves. *Arch Environ Health* 12:729-732, 1966.

20. Goldring IP, Cooper P, Ratner IM et al: Pulmonary effects of sulfur dioxide exposure in the Syrian hamster. *Arch Environ Health* 15:167-176, 1967.

21. Reid L: Experimental study of hypersecretion of mucus in the bronchial tree. *Brit J Exp Path* 44:437-455, 1964.

22. Greenwald I: Effects of inhalation of low concentrations of sulfur dioxide on man and other mammals. *Arch Industr Hyg Occup Med* 10:455-475, December 1954.

23. Goldring IP, Greenburg L, Park SS, et al: Pulmonary effects of sulfur dioxide exposure in the Syrian hamster: II. Combined with emphysema. *Arch Environ Health* 21:32-37, 1970.

24. Vaughn TR Jr, Jennelle LF, Lewis TR: Long-term exposure to low levels of air pollutants: Effects on pulmonary function in the beagle. *Arch Environ Health* 19:45-50, 1969.

25. Lewis TR, Campbell KI, Vaughn TR: Effects on canine pulmonary function via induced NO_2

impairment, particulate interaction, and subsequent SO$_x$. *Arch Environ Health* 18:596-601, 1969.

26. Alarie Y, Ulrich CE, Busey WM, et al: Effects of long-term continuous exposure of guinea pigs to sulfur dioxide. *Arch Environ Health* 21:769-777, 1970.

27. Threshold limit values of airborne contaminants for 1968. Read before the American Conference of Governmental Industrial Hygienists, (American Industrial Hygiene Association), 1968.

28. West P, Gaeke GC: Fixation of sulfur dioxide as disulfitomercurate (II) and subsequent colorimetric estimation. *Anal Chem* 28:1816-1819, 1956.

29. Alarie Y, Ulrich CE, Haddock RH, et al: Measurements of respiratory system flow resistance in cynomolgus monkeys and guinea pigs with digital computer techniques. *Arch Environ Health* 21: 483-491, 1970.

30. Alarie Y, Ulrich CE, Krumm A, et al: Measurements of the mechanical properties of the lung in cynomolgus monkeys with digital computerization. *Arch Environ Health* 22:643-654, 1971.

31. Alarie Y, Krumm A, Jennings HJ, et al: Measurement of the distribution of ventilation in cynomolgus monkeys with real-time digital computerization. *Arch Environ Health* 22:633-642, 1971.

32. Bates DV: The uptake of carbon monoxide in health and emphysema. *Clin Sci* 2:21-32, 1962.

33. Lewis BM, Lin TH, Noe FE, et al: The measurement of pulmonary diffusing capacity for carbon monoxide. *J Clin Invest* 38:2073-2086, 1959.

34. Kruhoffer P: Studies on the lung diffusion coefficient for carbon monoxide in normal subjects by means of C^{14}O. *Acta Physiol Scand* 32:106-111, 1954.

35. Long JE, Hatch TF: A method for assessing the physiological impairment produced by low level exposure to pulmonary irritants. *Amer Industr Hyg Asso J* 22:6-13, 1961.

36. Banerjee CM, Alarie Y, Woolard M: Gas tension in conscious monkeys. *Proc Soc Exp Biol Med* 128:1183-1185, 1968.

37. Page ES: Continuous inspection schemes. *Biometrika* 41:100-114, 1954.

38. Barnard GA: Control charts and stochastic processes. *J Roy Statist Soc*, series B. 21:240-271, 1959.

39. Woodward RH, Goldsmith PL: *Cumulative Sum Techniques*, Imperial Chemical Industries monograph 3. Edinburgh, Oliver & Boyd Ltd, 1964.

40. Murphy SD, Ulrich CE, Leng JK: Altered function in animals inhaling conjugated nitro-olefins. *Toxic Appl Pharmacol* 5:319-330, 1963.

41. Amdur MO: The respiratory response of guinea pigs to the inhalation of acetic acid vapor. *Amer Industr Hyg Assoc J* 22:1-5, 1961.

42. Murphy SD, Ulrich CE, Frankowitz SH, et al: Altered function in animals inhaling low concentrations of ozone and nitrogen dioxide. *Amer Industr Hyg Assoc J* 25: 246-253, 1964.

43. Rossing RG: Physiologic effects of chronic exposure to phosgene in dogs. *Amer J Physiol* 207:265-272, 1964.

44. DuBois, AB, Dautrebande L: Acute effects of breathing inert dust particles and of carbachol aerosol on the mechanical characteristics of the lung in man: Changes in response after inhaling sympathomimetic aerosols. *J Clin Invest* 37:1746-1755, 1958.

45. McDermott M: Acute respiratory effects of the inhalation of coal dust particles. *J Physiol* 162: 53P, 1962.

46. Cherniack RM: The physical properties of the lung in chronic obstructive pulmonary emphysema. *J Clin Invest* 35:394-404, 1956.

47. Butler J, Caro CG, Alcala R, et al: Physiological factors affecting airway resistance in normal subjects and in patients with obstructive respiratory disease. *J Clin Invest* 39:584-591, 1960.

48. Mead J, Lindgreen I, Gaensler EA: The mechanical properties of the lungs in emphysema. *J Clin Invest* 35:327-335, 1956.

49. Du Bois AB, Botelho SY, Comroe JH: A new method for measuring airway resistance in man using a body plethysmograph: Values in normal subjects and in patients with respiratory disease. *J Clin Invest*, 35:327-335, 1956.

50. Hogg JC, Macklem PT, Thurlbeck WM: Sites and nature of airway obstruction in chronic obstructive lung disease. *New Engl J Med* 278:1355-1360, 1968.

51. Bouhuys A: Distribution of inspired gas in the lung, in Field J (chief ed): *Handbook of Physiology: A Critical Comprehensive Presentation of Physiological Knowledge and Concepts*. Section 3: *Respiration*, vol 1. WO Fenn, H Rahn (section eds), Baltimore, Williams & Wilkins Co, 1964, pp 715-733.

52. Bouhuys A, Jönsson R, Lundin G: Non-uniformity of pulmonary ventilation in chronic diffuse obstructive emphysema. *Acta Med Scand* 162:29-46, 1958.

53. Colldahl H, Lundin G: Ventilatory studies of the lungs in asthma. *Acta Allerg* 5:37-51, 1952.

54. Bouhuys A, Lindell SE, Lundin G: Experimental studies on byssinosis. *Brit Med J* 1:324-326, 1960.

55. Bouhuys A, Jönsson R, Lichtneckert S, et al: Effects of histamine on pulmonary ventilation in man. *Clin Sci* 19:79-94, 1960.

56. Forster RE: Diffusion of gases, in Field J (chief ed): *Handbook of Physiology: A Critical Comprehensive Presentation of Physiological Knowledge and Concepts*. Section 3: *Respiration*, vol 1. WO Fenn, H Rahn (section eds), Baltimore, Williams & Wilkins Co, 1964, pp 839-872.

57. Forster RE: Interpretation of measurements of pulmonary diffusing capacity, in *Handbook of Physiology: A Critical, Comprehensive Presentation of Physiological Knowledge and Concepts*. Section 3: *Respiration*, vol 2. WO Fenn, H Rahn (section eds), Baltimore, Williams & Wilkins Co, 1965, pp 1453-1468.

58. Rinehart WE, Hatch TF: Concentration-time product (CT) as an expansion of dose in sublethal exposure to phosgene. *Amer Indust Hyg Assoc J* 25: 545-553, 1964.

59. Young WA, Shaw DB, Bates DV: Effects of low concentration of ozone on pulmonary function in man. *J Appl Physiol* 19:765-768, 1964.

60. Young WA, Shaw DB, Bates DV: Pulmonary function in welders exposed to ozone. *Arch Environ Health* 7:337-340, 1963.

61. MacFarland HN, Ulrich CE, Martin A, et al: Chronic exposure of cynamolgus monkeys to fly ash, in *Inhaled Particles and Vapours, III*. Oxford, England, Pergamon Press, to be published.

62. Amdur MO, Underhill, DW: The effects of various aerosols on the response of guinea pigs to sulfur dioxide. *Arch Environ Health* 16:460-468, 1968.

63. Amdur MO, Underhill DW: Response of guinea pigs to combination of sulfur dioxide and open hearth dust. *J Air Pollut Contr Assoc* 20:31-34, 1970.

60

Air pollution in nurseries:

Correlation with a decrease in

oxygen-carrying capacity of hemoglobin

Richard E. Behrman, David E. Fisher, and John Paton

T H E   E S T A B L I S H M E N T of special hospital units for newborn infants has contributed significantly to better medical care for these children. An important feature of these units is the artificial control of the circulating air in respect to temperature, humidity, and source; however, this regulation may itself involve new potential hazards. The environment of nurseries must be continually re-evaluated to avoid adverse fluctuations in temperature and humidity, the untoward effects of air contamination by microorganisms, inappropriate alterations in the intensity and duration of lighting, elevations in the level and frequency of noise, unnecessary

Supported by State of Illinois Department of Mental Health Grant 17-344, and Training Grant T-1-AM 5344 of National Institute of Arthritis and Metabolic Diseases, National Institutes of Health, United States Public Health Service.

restrictions on maternal contact with infants, and stray current from electrical equipment. An additional potential hazard may also result from the presence of inorganic particulate matter and/or gaseous pollutants circulated in the nursery air. The following report presents the results of a preliminary investigation of one aspect of this problem and emphasizes the need for further intensive investigation of the effects of air pollution on infants in nurseries.

METHODS

The relationship between the oxygen-carrying capacity of hemoglobin and the degree of air pollution in Chicago from September to December, 1969, was investigated in 25 newborn infants. All of these infants were in good health and weighed more than 2,500 Gm.; they were delivered and cared for in an air-conditioned nursery on the second floor (approximate elevation 27 feet above the street) of the University of Illinois Hospital which is located 1½ miles from the pollution station where the air samples were obtained. Both the station and the hospital are within one fourth of a mile of commuter expressways. Outside air is drawn into the nursery unit at the second floor level from an air-intake port facing an internal court with access to the street; the air is passed through 2 filters and an electrostatic field to eliminate particulate matter. A minimum of 25 per cent of the nursery air is from this outside source and the remainder is from general hospital air. There was no special source of air pollution identifiable in the vicinity of the intake. The peak daily carbon monoxide air concentrations were obtained from the pollution station during the period of the study; no monitoring instruments were available at the Medical Center at the time of this study. The daily peak carbon monoxide level in parts per million (ppm) was selected to represent an index of the general level of air contamination.

Heparinized venous blood was obtained each morning from infants in the nursery during their first 4 days of life. The oxygen capacity of hemoglobin was determined in duplicate on a Beckman GC-5 gas chromatograph. The concentration of oxygen used to calibrate this instrument was determined by the method of Scholander.[1] The method of determining the oxygen capacity has been previously reported.[2] The hemoglobin concentration was determined spectrophotometrically by the cyanmethoglobin method; hemoglobin concentrations were measured after the reaction had reached completion. The hemoglobin-oxygen capacities calculated from the hemoglobin concentrations with Hoeffner's coefficient were compared to those measured directly with the gas chromatograph. The differences between these values were then expressed as a percentage decrease from the calculated capacity. A decrease in the gas chromatographically measured oxygen-carrying capacity of greater than 2 per cent of the capacity calculated from the hemoglobin concentration exceeds the 95 per cent confidence limits; the discrepancy is due to differences in methodology. The percentage of carboxyhemoglobin was measured with an IL Co-oximeter.

RESULTS

The measured oxygen-carrying capacity in these infants was, with 2 exceptions, more than 2 per cent lower than the oxygen-carrying capacity of hemoglobin derived from the hemoglobin concentration. Table I demonstrates that when the environmental carbon monoxide level was greater than 20 ppm (high days) the decrease in the oxygen-carrying capacity was significantly greater than when the carbon monoxide level of air was 5 to 20 ppm (low days). At the same

Table I. Decrease (%) in oxygen-carrying capacity of hemoglobin

	Age (hr.)	
	0 - 24	*25 - 85*
Low	8.4 ± S.E. 1.11	11.0 ± S.E. 0.73
High	11.4 ± S.E. 1.08	13.9 ± S.E. 0.65
P value	< 0.025	< 0.005

High days were those when the environmental carbon monoxide level was greater than 20 ppm; low indicates those days when the carbon monoxide level in air was 5 to 20 ppm.

Table II. Comparison of carboxyhemoglobin concentration ([COHb]) and percentage decrease in oxygen-carrying capacity ($\downarrow$ Cap. O_2)

	[COHb]	$\downarrow$ Cap. O_2
	5.8	8.5
	9.8	14.6
	5.0	13.4
	9.6	13.7
	5.9	16.1
	9.4	13.4
	6.1	10.0
	5.5	11.1
	7.0	11.0
	5.4	13.6
	5.6	12.3
	6.6	15.8
	6.0	13.7
	6.0	14.6
	12.7	12.1
	5.3	17.0
Mean	6.98	13.18
S.E.	0.547	0.569

level of carbon monoxide (greater than 20 ppm) the older infants who had been breathing nursery air for the longest periods of time had a greater decrease in oxygen-carrying capacity (p < 0.025 in Fig. 1). The decrease in oxygen-carrying capacity in these infants did not correlate with a history of maternal smoking or smoking after admission during labor or with the concentrations of serum bilirubin in these infants. There was no evidence to suggest abnormal rates of hemolysis in any of these infants. The mean percentage concentration of carboxyhemoglobin was 6.98 ± S.E. 0.55 in 16 of these babies in whom it was measured; this was significantly less than their percentage of decrease in oxygen-carrying capacity ($p < 0.001$) and did not correlate with the atmospheric carbon monoxide (Table II).

DISCUSSION

The large range of values (Fig. 1) obtained in this survey reflects the inaccuracy of using air pollution measurements obtained some distance from the nurseries which could not take into account: (1) variations in the level of carbon monoxide and other pollutants due to local patterns of air flow (inversion, etc.), (2) differences in the level of elevation (floor) between where air was obtained for analysis and the location of the patients in the hospital, (3) the quantity of automobile traffic in the vicinity of the University Hospital, and (4) variations in the composition of hospital air on different days. It also reflects the limitation of using carbon monoxide as the only index of air pollution in this study. In addition, an analysis of the average hourly carbon monoxide levels during a 2 week period in November showed that the peak values were only a rough reflection of the daily pattern of carbon monoxide concentrations in the air. Considering these limitations, the fact that a significant relationship was found between the level of one air pollutant, carbon monoxide, and a decrease in the oxygen-carrying capacity of infants is perhaps even more remarkable. In order to evaluate these variants we are presently studying the concentrations of a number of air pollutants in the nursery and in incubators as factors which may affect the oxygenation of newborn infants.

The cumulative errors in our determina-

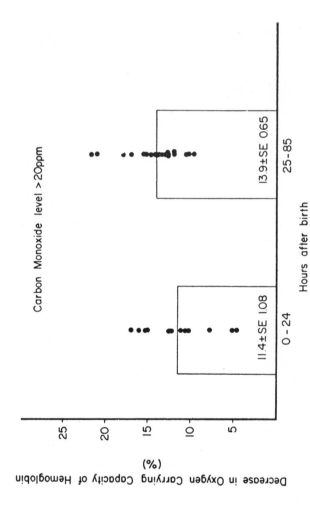

Fig. 1. Percentage decrease in oxygen-carrying capacity of hemoglobin at high levels of air pollution plotted against age of infants.

tions of the oxygen-carrying capacities and hemoglobin concentrations were substantially less than the discrepancy between the measured and hemoglobin derived oxygen-carrying capacities. Variations in hemoglobin concentrations over the range of concentrations increase the error in the gas chromatographic found in these patients does not significantly analysis.[1, 3]

The concentrations of carboxyhemoglobin in the blood of normal nonsmoking mothers and their fetuses has been reported to vary from 0.4 to 2.6 per cent and 0.7 to 2.5 per cent, respectively; in smoking mothers and their fetuses the levels reported range from 2.0 to 8.3 per cent and 2.4 to 7.6 per cent, respectively.[4] Normal values during the first week of life are generally less than 1 per cent carboxyhemoglobin in infants of nonsmoking mothers cared for in what was probably a low-pollution environment.[4, 5] The carboxyhemoglobin values presented in this report are above the levels expected to be found in infants of nonsmoking mothers. This may be a result of the inadequacies of the histories we obtained from our mothers who were predominantly from a low socioeconomic level. Similarly, these carboxyhemoglobin concentrations are greater than those usually obtained in infants with hemolytic disease.[4, 6] However, the levels reported in this study are consistent with the carboxyhemoglobin concentrations that can be calculated to occur in newborn infants whose nonsmoking mothers are assumed to have equilibrated with relatively high environmental levels of carbon monoxide.[4] For example, if a mother had equilibrated with an atmospheric concentration of 20 ppm, the infant's carboxyhemoglobin concentrations at birth can be estimated at 4.8 per cent; at 40 ppm ambient carbon monoxide the newborn infant's carboxyhemoglobin would be 8.8 per cent.

Inhalation by the infants in this study of

carbon monoxide in the concentrations reported (4 to 35 ppm), coupled with possible placental transfer of carbon monoxide from maternal blood[4, 7] and with endogenous carbon monoxide production in the infants,[5] cannot account for all of the observed decrease in oxygen-carrying capacity of hemoglobin. There is the unexplored possibility that other atmospheric pollutants (sulfur or nitrogenous gaseous compounds in particular) or their in vivo metabolites may combine with hemoglobin to produce methemoglobin, sulfhemoglobin, or other nonoxygen-carrying hemoglobin compounds that could contribute to a decrease in the oxygen-carrying capacity. This possibility is now under investigation. Therefore, at present the level of environmental carbon monoxide should only be considered a reflection of the general degree of air pollution rather than the cause of all of the observed decrease in oxygen-carrying capacity.

No untoward clinical effects from air pollution were observed in these infants. The decreases in hemoglobin-oxygen–carrying capacity represent a relatively small functional anemia which is probably of little significance to the normal newborn infant who has a relatively high hemoglobin concentration compared to the adult. However, the measured concentrations of carboxyhemoglobin could potentially result in a slight increase in the oxygen affinity of fetal hemoglobin (2 to 4 mm. Hg at arterial pH 7.35 to 7.40 and Pco_2 of 40 mm. Hg) with a consequent decrease in the amount of oxygen that can be unloaded in the tissues at a given oxygen tension when compared to the normal neonate's oxygen affinity.[8-11, 14] Small amounts of carboxyhemoglobin (4.95 to 9.69 per cent) can result in decreased oxygen tensions in arterial and mixed venous blood.[12] The effects of other gaseous pollutants on the shape and position of the hemoglobin-oxygen dissociation curve are unknown. Thus the combination of hemoglobin with carbon monoxide and/or some other air pollutant could potentially magnify the effect of physiologic venoarterial shunting in the lungs[13] or impose an abnormal resistance to diffusion of oxygen within the blood.[14] This may be equivalent to marked reductions in tissue blood flow or in blood concentration of hemoglobin[15] and requires a compensatory increase in cardiac output. These effects may be exaggerated in the presence of hypoxia by increasing the ratio of carboxymyglobin to carboxyhemoglobin in the myocardium.[4, 16] There is the remote possibility of an atmospheric pollutant interfering with the function of enzymes.[4, 17, 18] Thus even small decreases in oxygen-carrying capacity might further impair the tissue uptake delivery of oxygen and possibly increase hypoxia in newborn infants with respiratory or cardiac disease.

Our purpose in this preliminary report is to call attention to a potential pediatric health problem. These observations suggest the need to study the pattern of air pollution in different patient areas within a hospital[10] and the effects of a variety of pollutants, including carbon monoxide, on the transport of oxygen in blood.[19] Nurseries and areas used for neonatal intensive care centers have often been designed to circulate a large proportion of outside air rather than internal hospital-filtered air owing to the risk of disseminating antibiotic-resistant hospital organisms; the circulation of outside air now also should be considered a possible potential environmental hazard for the neonate in the hospital.

REFERENCES

1. Scholander, P. F.: Analyzer for accurate estimation of respiratory gases in one-half cubic centimeter samples, J. Biol. Chem. **167**: 235, 1947.

2. Behrman, R. E.: In vivo oxygen curves for maternal and fetal rhesus monkeys, J. Appl. Physiol. 25: 224, 1968.
3. Paton, J. B., Peterson, E., Fisher, D., and Behrman, R. E.: Oxygen dissociation curves of fetal and adult baboons. In press.
4. Longo, L. D.: Carbon monoxide in the pregnant mother and fetus, and its exchange across the placenta. In press.
5. Fallstrom, S. P.: On the endogenous formation of carbon monoxide in full-term newborn infants, Acta Paediat. Scand. (Suppl.) 189: 1, 1969.
6. Wranne, L.: Studies on erythro-kinetics in infancy. XIV. The relation between anemia and hemoglobin catabolism in Rh-hemolytic disease of the newborn, Acta Paediat. Scand. 58: 49, 1969.
7. Haddon, W., Jr., Nesbitt, R. E. L., and Garcia, R.: Smoking and pregnancy: Carbon monoxide in blood during gestation and at term, Obstet. Gynec. 18: 3, 1961.
8. Douglas, C. G., Haldane, J. S., and Haldane, J. B. S.: The laws of combination and haemoglobin with carbon monoxide and oxygen, J. Physiol. 22: 231, 1897.
9. Roughton, F. J. W., and Darling, R. C.: The effect of carbon monoxide on the oxyhemoglobin dissociation curve, Amer. J. Physiol. 141: 17, 1944.
10. Coburn, R. F., Forster, R. E., and Kane, P. B.: Considerations of the physiological variables that determine the blood carboxyhemoglobin concentration in man, J. Clin. Invest. 44: 1899, 1965.
11. Brody, J. S., and Coburn, R. F.: Carbon monoxide-induced arterial hypoxemia, Science 164: 1297, 1969.
12. Ayres, S. M., Giannelli, S., Jr., and Armstrong, R. G.: Carboxyhemoglobin: hemodynamic and respiratory response to small concentrations, Science 149: 193, 1965.
13. Ayres, S. M., Criscitiello, A. M., and Grabovsky, E.: Components of alveolar-arterial O_2 difference in normal man, J. Appl. Physiol. 19: 43, 1964.
14. Roughton, F. J. W., and Forster, R. E.: Relative importance of diffusion and chemical reaction rates in determining rate of exchange of gases in the human lung, with special reference to true diffusing capacity of pulmonary membrane and volume of blood in the lung capillaries, J. Appl. Physiol. 11: 290, 1957.
15. Permutt, S., and Farhi, L.: Tissue hypoxia and carbon monoxide, in effects of chronic exposure to low levels of carbon monoxide in human health, behavior and performance,

Nat. Acad. Sci. and Nat. Acad. Engineering, Washington, D. C., 1969, pp. 18-24.

16. Coburn, R. F.: The carbon monoxide body stores. In press.

17. Ball, E. G., Strittmatter, C. F., and Cooper, O.: The fraction of cytochrome oxidase with carbon monoxide, J. Biol. Chem. **193:** 635, 1951.

18. Meldrum, N. U., and Roughton, J. W.: Carbonic anhydrase. Its preparation and properties, J. Physiol. **80:** 113, 1933.

19. Committee on effects of atmospheric contaminants on human health and welfare in reports on effects of chronic exposure of low levels of carbon monoxide on human health, behavior and performance, National Academy of Science, 1969.

ESTABLISHMENT OF "THRESHOLD" CO EXPOSURE LEVELS*

Arthur B. DuBois, M.D.†

This paper touches on four topics. First, I shall consider the CO levels that have already been established for various purposes, next, how levels to be proposed for continuous exposure of the population at large must differ from those presently existing for exposure of special groups over limited periods of time, and after that, examine the question whether a "threshold" exists for CO exposure. Finally, we may review what steps remain to be taken during the establishment of air quality standards for CO.

Present Standards

The presently recommended limits for CO exposure of special groups for limited duration are listed in TABLE 1. This Table does not include values that would be acceptable for continuous exposure of the population at large throughout a lifetime, inasmuch as such standards are not yet available.

The first three lines of TABLE 1 list the short-term military and space vehicle standards for CO. The criteria used in arrival at such standards have been described.[1,2] These standards were not believed to be suitable for use in normal operation, but were only acceptable for forseeable emergencies. The emergency levels for CO recommended by the Committee on Toxicology of the National Research Council to military and space agencies were 1,500 ppm for ten minutes, 800 ppm for 30 minutes, or 400 ppm for 60 minutes.[3] When mental acuity is required, the corresponding CO levels should be 1,000, 500 and 200 ppm.[4] In addition, the Navy adopted recommendations of the NRC Committee on Toxicology[5] of an emergency limit of 200 ppm for 24 hours in the event of a spill of CO. The criterion used in selection of this level was that it should permit some discomfort and minimal toxic changes believed to be readily reversible. Exposure to these emergency levels was not to be repeated until after recovery from the effects of a preceding exposure. The population under consideration consisted of young, healthy adults.

Industrial Limits

The industrial threshold limit value (TLV) adopted by the American Conference of Governmental Industrial Hygienists in 1966[6] as a time-weighted average limit was 50 ppm (8–10% COHb) assuming a population of industrial workers exposed seven to eight hours a workday and for a 40-hour workweek. The Maximal Acceptable Concentrations (MAC's) for materials encountered in industrial atmospheres, set by the U.S.A. Standards Institute, gives an eight-hour ceiling value (not time-weighted average) of 100 ppm.[7] The USSR[8] sets this MAC at 18 ppm (20 mg/M^3).

Continuous Exposures for Military and Space Vehicles

The Navy adopted a 25 ppm limit (4–5% COHb) for 90-day submarine exposure of a young, healthy, adult population under medical surveillance.[5] The

* This work was supported in part by a research grant, HE 4797, from the Public Health Service.

TABLE 1
CARBON MONOXIDE LEVELS WHICH HAVE BEEN RECOMMENDED IN THE PAST

ppm	COHb%	Duration	Exposure Conditions	Subjects	Purpose	Source
1,500		10 min	EEL* normal activity	healthy young adults	military & space	NRC
1,000		10 min	EEL mental acuity	healthy young adults	military & space	NRC
800		30 min	EEL normal activity	healthy young adults	military & space	NRC
500		30 min	EEL mental acuity	healthy young adults	military & space	NRC
400		60 min	EEL normal activity	healthy young adults	military & space	NRC
200		60 min	EEL mental acuity	healthy young adults	military & space	NRC
200		24 hr	EEL normal activity	healthy young adults	military & space	NRC
100		8 hr	MAC†	industrial workers	industry	USASI
18		8 hr	MAC	industrial workers	industry	USSR
50	8–10	7–8 hr/d	TLV‡ 40 hr/wk	industrial workers	industry	ACGIH
25	4–5	90 d	continuous	healthy young adults	nuclear submarine	NRC
15(T)§	2–3	90–1,000 d	continuous	healthy young adults	spacecraft	NRC
?¶		lifetime	continuous	population at large	ambient air	NRC

* EEL = "emergency exposure limit."
† MAC = "maximal acceptable concentration."
‡ TLV = "threshold limit value," time weighted average.
§ T = a provisional level for continuous exposure, 12-hour integrated time average.
¶ ? = a level which has not yet been established.

73

provisional limit for 90- and 1,000-day continuous exposure in United States space cabin atmospheres is 15 ppm (15 mmoles per 25 cubic meters), based on an integrated exposure over a 12-hour period, and this would yield 2 to 3% COHb. This limit was recommended partly on the grounds that CO at higher levels might compromise the high level of judgment and performance required of pilots and other occupants of space vehicles.[3]

Population at Large

So far, we have been discussing CO levels or limits applicable to an industrial working population, military personnel, or space pilots. In considering public health limits, we have to include a broader spectrum of people exposed. These range from the as yet unborn child and expectant mother through the infant, adult, elderly person and invalid to the hospitalized patient barely clinging to life. Would CO affect these individuals more than it would affect a factory worker? Since CO acts by encroachment on tissue oxygen tension, it is logical to expect that it would affect individuals who have a low or limited tissue oxygen. Such individuals are those who have anemia, pulmonary insufficiency, or cardiovascular disease.[9]

The performance of certain tasks may be impaired by CO. These include those that require the use of special senses, such as night vision, or of timing, judgment, mental agility, or heavy exercise.

Contributing to, and enhancing, the effects of inhaled ambient CO are exposed to altitude, and to inhaled cigarette smoke, which itself contains CO. Some toxic agents, such as nitric oxide, by affecting blood hemoglobin, theoretically can enhance the effect of CO.

Is There a "Threshold" for CO Effects?

We have to ask whether there is any really safe level of CO. If there were, one would look for a "threshold," or level below which no adverse effects occurred. However, three studies on neurological function have shown a response to CO proportional to the CO concentration and extending straight back toward zero disturbance at zero concentration in the inspired air, but proportional on upward from zero. These studies, cited in reference 9, are those of Mills and Edwards on the carotid body, MacFarland and coworkers, on the retina, and Beard and Wertheim on discrimination of tone duration. It seems that CO does not have a "hockey-stick" dose-response curve—one that would be initially flat, reach a threshold, and then rise—but instead, a line or curve that passes through the origin of axes. We should then rephrase the title of this paper to ask: How much decrement in function are we willing to tolerate under conditions of public exposure for a specified time at a given concentration profile and in specified locations?

Steps Leading toward Standards for CO

As one step toward the establishment of air quality standards, the National Air Pollution Control Administration has prepared a document on Air Quality Criteria for CO.[10] Each state must select air quality standards for CO. These standards will require coordination to make them compatible with the air shed regions, and to provide industry with guidelines such that the products (motor vehicles) will be in accordance with the levels adopted by different states. The last line in TABLE 1 has been left partly blank to allow the reader to fill in the appropriate information once it has become available.

74

References

1. SMYTH, JR., H. F. *et al.* 1964. Basis for establishing emergency inhalation exposure limits applicable to military and space chemicals. Committee on Toxicology, NAS–NRC. Washington, D. C.
2. SMITH, JR., H. F. 1966. Military and space short-term inhalation standards. Arch. Environ. Health 12: 488–490.
3. SPACE SCIENCE BOARD, NAS. 1968. Atmospheric contaminants in spacecraft. Report of the Panel on Air Standards for Manned Space Flight. 2nd printing, Washington, D. C.
4. NATIONAL RESEARCH COUNCIL'S COMMITTEE ON TOXICOLOGY. 1969. Supplementary note.
5. NATIONAL ACADEMY OF SCIENCES. A compendium of recommendations for safe concentrations of atmospheric contaminants. Contract Nonr 2300(29), NR 303–366. Department of the Navy Office of Naval Research.
6. AMERICAN CONFERENCE OF GOVERNMENTAL INDUSTRIAL HYGIENISTS. 1966. Threshold limit values of airborne contaminants. Cincinnati, Ohio.
7. UNITED STATES OF AMERICA STANDARDS INSTITUTE. 1965. Irish, D. D. Arch. Environ. Health 10: 546–549: Magazine of Standards. 1967. : 47–49.
8. RUSSIAN MAC'S. 1959. Translation. Industrial Hygiene and Occupational Disease. No. 5. : 7–15.
9. DIVISION OF MEDICAL SCIENCES, NAS–NAE. 1969. Effects of chronic exposure to low levels of carbon monoxide on human health, behavior, and performance. Committee on Effects of Atmospheric Contaminants òn Human Health and Welfare. Washington, D.C.
10. UNITED STATES DEPARTMENT OF HEALTH, EDUCATION, AND WELFARE. 1969. Air quality criteria for carbon monoxide. Public Health Service. Consumer Protection and Environmental Health Service. National Air Pollution Control Administration. Bureau of Criteria and Standards. Durham, N. C. In preparation.

Toxicity of Particulate Air Pollutants

TUMORS IN MICE INDUCED
BY AIR PARTICULATE MATTER
FROM A PETROCHEMICAL INDUSTRIAL AREA*

R. H. RIGDON AND JACK NEAL

Tumors; pollutants, air; petrochemical area; mice

ABSTRACT

Air pollutants have been collected day and night, 6 days a week, for 5 years at a single station in an area where there are several petrochemical plants. A marked increase occurred in the total particulate matter per cubic meter of air, milligrams of benzene-solubles per 1000 cubic meters of air, and in the amount of benzo(a)-pyrene per 1000 cubic meters of air between 1965 and 1969.

The benzene-solubles prepared from these air samples and injected subcutaneously into mice produced local fibrosarcomas. Benzene-solubles were obtained from air sampled at 6 other stations elsewhere in the country during this same period. No tumors occurred when these benzene-solubles were injected subcutaneously into the same strain of mice.

The amount of benzo(a)pyrene in the benzene-solubles varied from month to month and year to year. The number of subcutaneous fibrosarcomas, however, was less in the mice given benzo(a)pyrene (Nutritional Biochemical Co.) than in the same strain of mice injected subcutaneously with corresponding amounts of the benzo(a)pyrene in the benzene-solubles. This variation may be explained on the basis that either carcinogens or cocarcinogenic agents, other than benzo(a)pyrene, also were present in the benzene-solubles obtained from the air particulate matter.

The demonstration of tumors in the subcutaneous tissue of mice following the local injection of benzene-solubles obtained from particulate matter collected from the air cannot be applied theoretically to man. However, such observations may be considered as the first link in the chain leading to determination of the carcinogenic effect of air pollutants in the lung of man.

Hueper et al. (1) in 1962 stated that the information available indicates that an effective and rational control of lung cancer depends on the availability of a greatly increased, and more specific, knowledge

* Supported by Grant DHEW 5 RO1 CA 08147–05 from the National Cancer Institute, National Institutes of Health, USPHS.

78

of the nature, quantity, physical status, and exposure mechanism of specific carcinogenic factors, as well as of the presence and action of any associated promoting cocarcinogenic and anticarcinogenic pollutants and climate conditions present in a particular community. In their study of air pollutants from 8 American cities, the benzo(a)-pyrene content was determined and the carcinogenic potency tested by the subcutaneous injection into mice.

Sawicki (4) in 1967 listed many carcinogenic compounds present in the atmosphere of different cities and gave the average amount of some. Those represented included n-alkanes, proteins, aromatic hydrocarbon, azo and imino heterocyclic compounds, polynuclear carbonyl compounds, quinones, acids, and phenols. Sawicki noted that one of the most active carcinogens in the urban atmosphere is benzo(a)-pyrene and that an increase may also be found in the concentrations of other polynuclear aromatic hydrocarbons when the concentration of benzo(a)pyrene is high 4.

Much effort has been spent in determining the concentration of benzo(a)pyrene in the air in view of the above relationship. The benzo(a)pyrene content of urban atmosphere apparently was first reported in 1952 by Waller (10) and in 1954 by Kotin et al. (2). The concentration of benzo(a)pyrene in the atmosphere of highly industrialized European cities is definitely greater than that of the larger cities in America (4).

We have collected airborne particulate matter day and night, 6 days a week, for 5 years from a petrochemical industrial area in Texas and determined that matter's benzo(a)pyrene content. The benzene-soluble fraction has been injected subcutaneously in mice, a high percentage of which developed a local fibrosarcoma. Our experimental study follows.

MATERIALS AND METHODS

A high volume air sampler, capacity 50 c.f.m., with type A glass fiber filter, manufactured by Gelman Instrument Co., Ann Arbor, Mich., was used. The particulate matter was collected from the air day and night, 6 days a week, for 5 years. The collecting station was located in Texas City near petrochemical plants, the chief products of which are styrene monomer, vinyl chloride monomer, acrylonitrile monomer, polyethylene polymer, methonal tertiarybutylamine, detergent bases, resins of all types, high octane gasoline, and carbon black. Inorganic chemicals, such as tin, sulphuric acid, arsenic, dicalcium phosphate, and fertilizers, also are produced in this area. Control air samples were obtained by a similar technique from other areas in the state.

The air pollutants on the filters were brought to a constant weight by heating at 60 C before and after sampling. The pollutant was extracted with double–distilled benzene for 6 hours in a Soxhlet extractor. The benzene was evaporated at 60 C, and the samples were brought to a constant weight. The benzo(a)pyrene present in these extracts was determined quantitatively by the method used by Stanley et al. (9) and Sawicki et al. (8). The average yearly chemical assay data for the air samples collected between 1965 and 1969 are shown in Table 1. The monthly data for 1966, 1967, and 1969 are shown in Table 2. This average is based upon 12 to 14, 48-hour, total particulate air samples per month.

The benzene-soluble was prepared from air samples collected over varying periods of time and injected subcutaneously in a strain of CFW mice. The amount injected per mouse varied from 1 to 20 mg. This benzene-soluble was mixed thoroughly with cottonseed oil, and 0.5 ml were injected subcutaneously. Benzo(a)pyrene (Nutritional Biochemical Co., Cleveland, Ohio), lamp black (Monsanto Chemical Co.), carbon black (Dr. Carl A. Nau, University of Oklahoma Medical Center, Oklahoma City, Okla.), and cottonseed oil (Wesson Oil) were injected as controls and to allow comparison of the frequency and type of tumors with those resulting from the benzene-soluble prepared from the air pollutant. The amount of each carcinogenic agent injected subcutaneously and the number of mice in each experiment are given in Table 3. Benzene-soluble was prepared from the lamp black, utilizing the same technique as that used for the air pollutant.

The mice were an inbred strain of CFW white Swiss used in this laboratory for the past 10 years. They were usually 30 to 50 days old when injected subcutaneously and were observed frequently for a year. They were housed in small groups and fed Purina laboratory pellets *ad libitum*. When a tumor was observed the animal was killed, and the tumor was removed and fixed in a 10% solution of formalin. Paraffin sections were prepared and stained routinely with hematoxylin and eosin. The tumors that developed in 4 of the mice were transplanted in other mice.

RESULTS

The average number of cubic meters of air sampled in each of the years between 1965 and 1969 is shown in Table 1. There is a progressive increase in the yearly average of particulate matter collected during the first 4 years. There is also a corresponding increase in the total particulate matter, the benzene-soluble, the percentage of ben-

TABLE 1

Air samples—chemical assay data—yearly average

Year	Air sampled (m³)	Total particulate matter (μg/m³)	Benzene-solubles (mg/1,000 m³)	% Benzene solubles in particulate matter	Benzo(a)pyrene (μg/1,000 m³)
1965	12,000	61	3.46	3.1	1.02
1966	23,000	66	3.36	4.2	1.27
1967	43,000	77	3.70	4.9	1.50
1968	83,000	108	6.75	9.6	2.68
1969	82,000	110	6.90	9.5	2.69

80

TABLE 2

Chemical assay data—Texas City, Texas

Month	Total particulate matter of air sampled (µg/m³)			Benzene solubles of air sampled (mg/1,000 m³)			Percent benzene solubles in total particulate			Benzo(a)pyrene of air sampled (µg/1,000 m³)		
	1966	1967	1969	1966	1967	1969	1966	1967	1969	1966	1967	1969
January	58	62	60	2.49	2.56	5.6	3.1	3.5	7.75	0.8	1.2	1.8
February	82	93	62	2.70	2.78	5.7	4.0	5.2	8.20	1.1	1.4	2.1
March	73	90	50	4.05	4.25	4.2	5.0	6.0	7.30	1.5	1.7	1.7
April	62	74	112	4.71	6.00	4.9	5.0	6.2	8.65	1.6	2.3	1.7
May	73	94	70	3.40	3.02	5.8	3.9	4.0	11.80	1.3	1.5	2.2
June	56	63	89	3.10	3.25	3.5	5.7	6.3	8.60	1.2	1.4	1.4
July	42	80	102	2.17	2.50	5.9	3.4	3.2	7.80	1.3	1.1	2.3
August	60	74	111	2.95	2.80	6.7	3.5	4.1	9.00	1.2	1.3	2.8
September	69	79	170	3.34	5.10	10.0	4.3	5.3	12.20	1.4	1.8	3.7
October	80	86	180	3.76	4.90	11.2	4.2	4.8	13.60	1.5	1.7	4.2
November	72	71	160	3.95	3.90	9.5	5.0	5.3	11.00	1.4	1.6	4.0
December	66	54	155	3.66	3.40	8.0	3.4	4.8	10.10	1.2	1.4	3.9
Average	66	77	110	3.36	3.70	6.75	4.2	4.9	9.65	1.27	1.5	2.68

TABLE 3

Local tumors in mice following subcutaneous injections of control and carcinogenic agents

Subcutaneous injection	Amount injected		Number of mice	% with tumors	Interval in days between injection and death. Number with● and without° fibrosarcomas								
					0–50	50–75	75–100	100–150	150–200	200–250	250–300	300–350	350+
Cotton seed oil	0.5	ml	47	0							o–4	o–1	o–42
Carbon black	17.0	mg	49	2									●–1 o–48
Lamp black	8.0	mg	50	2									●–1 o–49
Benzene soluble of lamp black	10.0	mg	35	31	o–1	o–4	●–1 o–1	o–1	●–4 o–4	●–4 o–2	●–2 o–4	o–2	o–7
Benzo(a)pyrene	0.001	mg	18	0	o–1		o–1	o–1	o–1	o–1			o–13
Benzo(a)pyrene	0.01	mg	16	0		o–1		o–1	o–1		o–1	o–1	o–14
Benzo(a)pyrene	0.025	mg	15	33	o–1			●–2 o–1	●–3		●–1		o–9
Benzo(a)pyrene	0.05	mg	13	70	o–3	●–1	●–1	●–4 o–2	●–2 o–1	●–1 o–1			
Benzo(a)pyrene	0.1	mg	52	56	o–3		●–16 o–2	●–13 o–2	o–9	●–5 o–1			
Benzo(a)pyrene	0.5	mg	9	77	o–1		●–6	●–1 o–7					
Benzo(a)pyrene	1.0	mg	20	40				●–7 o–7	o–5				
Benzo(a)pyrene	2.0	mg	20	90		●–1 o–2	●–2	●–14 o–1	●–2				
Benzene-soluble	2.5	mg	50	40		o–1 o–2		●–1 o–5	●–2 o–2	●–2 o–2		o–1	o–23
Benzene-soluble	5.0	mg	54	61				●–11 o–2	●–8 o–2	●–4	●–4 o–1	●–2 o–2	●–14
Benzene-soluble	10.0	mg	46	50		o–1	●–2	●–3	●–6 o–2	●–9	●–5	o–3	o–17

82

TABLE 4

Local tumors in mice following subcutaneous injections of benzene-solubles obtained from air pollutants in Texas City from 1965 to 1969

Date injected	mg Benzene-soluble injected	µg Benzo(a)pyrene in benzene-soluble*	Number of mice	Number with tumor	%
11–24–65	1.0	0.29	20	0	..
11–24–65	10.0	2.94	20	1	5
2–15–66	5.0	1.85	24	0	..
3–23–66	5.0	1.85	14	1	7
12–13–66	6.0	2.34	4	0	..
12–13–66	10.0	3.90	40	0	..
4–19–67	5.0	2.10	10	0	..
9– 8–67	10.0	4.20	7	0	..
12–19–67	10.0	4.20	61	0	..
9–16–68	10.0	3.99	69	2	3
1– 3–69	10.0	3.90	23	12	52
5–16–69	10.0	3.90	59	7	12
10– 2–69	2.5	0.97	50	20	40
10– 2–69	5.0	1.85	54	33	61
10– 2–69	10.0	3.90	46	23	50

* Calculated from chemical assay data given in Tables 1 and 2.

zene-solubles, and the benzo(a)pyrene in these samples. There is essentially no difference, however, in the chemical assays of the air samples for the years 1968 and 1969. The chemical assay averages for each month for 1966, 1967, and 1969 are given in Table 2. There are some variations from month to month that might be influenced by weather conditions and production activity.

In Table 4 the data on the subcutaneous injection of the benzene-soluble prepared from the particulate matter are given with the amount, number of mice per experiment, and the frequency of occurrence of fibrosarcomas. A definite increase in the number of tumors first was observed in mice injected on Jan. 3, 1969, with the particulate matter collected during the latter part of 1968. In 1965 the percent of benzene-solubles in the total particulate matter and the benzo(a)-pyrene in the air samples were 3.1% and 1.02 µg per 1000 m³; in 1967 they were 4.9% and 1.50 µg, and in 1969 they were 9.65% and 2.68 µg. It is obvious, therefore, that the increase in frequency of fibrosarcomas in the subcutaneous tissues of the mouse is related to the increase of benzene-solubles and benzo(a)pyrene in the particulate matter. The amount of benzene-soluble (9.5%) and benzo(a)pyrene (2.69 µg per 1000 m³) in the particulate matter for 1969 is essentially the same as that for 1968 (9.65% benzene-soluble and 2.68 µg benzo-

Table 5

Benzo(a)pyrene content (µg per g) of benzene-solubles for 1965–1969*

Month	1965	1966	1967	1968	1969
January	. . .	320	470	380	362
February	. . .	410	502	402	380
March	. . .	370	398	400	405
April	. . .	320	382	380	370
May	. . .	380	490	410	395
June	. . .	388	430	415	400
July	330	600	440	398	395
August	246	420	460	425	420
September	400	420	350	380	370
October	314	390	347	395	390
November	382	356	410	385	388
December	250	330	410	310	315
Average	320	390	420	399	386

* Average of 12–14, 48-hour, high volume, total particulate matter air samples per month.

(a) pyrene per 1000 m³) (Table 1), but the frequency of fibrosarcomas is greater in 1969 than in 1968 (Table 4).

The benzo(a)pyrene content of the benzene-solubles for each month from 1965 to 1969 is given in Table 5. There are variations. However, these are not related to specific seasons of the year. In August and September, 1966, there were 420 µg per g of the benzene-solubles; in January and February, 1967, there were 470 and 502 µg per g; and in May and June, 1969, there were 395 and 400 µg per g. The averages for the years 1965 to 1969 were 320, 390, 420, 399, and 386 µg per g of the benzene-solubles (Table 5).

In October, 1969, the benzene-soluble was divided into 3 concentrations: 2.5 mg, 5.0 mg, and 10.0 mg per 0.5 ml of cottonseed oil (Table 4). The percentage of tumors was 40, 61, and 50, respectively. There is some variation in the frequency of tumors that is related apparently to the biological response of the mice since these 3 concentrations were obtained from the same collection of air pollutant.

The air pollutant samples obtained from 6 areas other than Texas City from 1966 and 1969 are shown in Table 6. Benzene-soluble was prepared from these samples and injected subcutaneously into mice of the same strain. A few tumors occurred in the mice given the benzene-soluble obtained from the station on the campus of the University of Texas Medical Branch, Galveston, Texas which is approximately 9 miles south of the sampling station in Texas City (Table 6). Additional control data for this study are included in Table 3. None

TABLE 6

Local tumors in mice following subcutaneous injection of benzene solubles obtained from air pollutants from 1966 to 1969

Site	Date injected	Amount injection/mg	Number of mice	Number of tumors	%
Corpus Christi, Texas	12–13–66	10	10	0	..
Corpus Christi, Texas	9–.8–67	10	18	0	..
Beaumont, Texas	12–13–66	10	10	0	..
Beaumont, Texas	9– 8–67	10	5	0	..
San Antonio, Texas	4-24-68	10	32	0	..
Odessa, Texas	4-24-68	10	62	0	..
North Carolina	1- 3-69	10	20	0	..
UTMB*	5-16-69	10	50	0	..
UTMB	5-16-69	20	50	3	6
UTMB	6-19-69	5	50	1	2
UTMB	6-19-69	2.5	52	0	..

* University of Texas Medical Branch, Galveston, Texas

of the 47 mice given 0.5 ml of the cottonseed oil developed a subcutaneous tumor. One mouse in a group of 49 given carbon black and 1 in a group of 50 given lamp black developed a fibrosarcoma. However, 31% of 35 mice given 10.0 mg of the benzene-soluble obtained from the same sample of lamp black did develop fibrosarcomas (Table 3).

The frequency of subcutaneous fibrosarcomas in the mice injected with different concentrations of benzo(a)pyrene is shown in Table 3. There is an increase in the number of tumors which accompanies an increase in benzo(a)pyrene injected. This increase, however, is not progressive. Fibrosarcomas developed in 90% of a group of 20 mice given 2.0 mg of benzo(a)pyrene and in 33% of 15 mice given 0.025 mg. None of the 18 mice given 0.001 mg or the 16 mice given 0.01 mg of benzo(a)pyrene developed a local tumor. The interval between the subcutaneous injection of the benzo(a)pyrene and the occurrence of the tumor is shown in Table 3. A similar observation was made on the time of occurrence of tumors in mice given the benzene-soluble obtained from the polluted air samples. This was similar to that in the group given only the benzo(a)pyrene.

Mice injected subcutaneously with the benzene-soluble prepared from the air pollutant frequently developed a cyst at the site of injection (Fig. 1). These cysts sometimes measured as much as 1.0 cm in diameter and were filled with a thin, brown, oily liquid. A minimal fibrous tissue wall surrounded these cysts. A neoplastic tumor some-

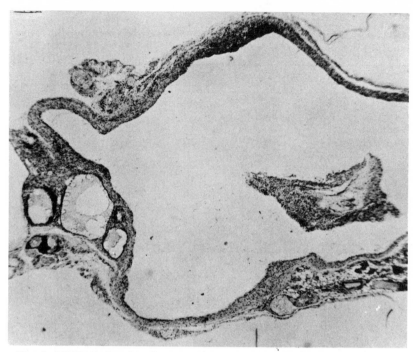

F<small>IG</small>. 1. M145–11. Cyst in subcutaneous tissue of a mouse injected 368 days previously with 10 mg of benzene-soluble obtained from the University of Texas Medical Branch station. H&E stain, X 25.

times developed in the wall of a cyst (Figs. 2,3). Other mice developed a fibrosarcoma at the injection site without development of any macroscopic cyst. Microscopic cysts, varying markedly in size, were present in some of the neoplastic lesions (Fig. 4). Some of the mice did not develop either a cystic or neoplastic lesion at the site of the subcutaneous injection of the benzene-soluble, but did have varying amounts of pigment with no reaction. In the earlier part of this study, several mice were killed when a local mass was observed at the site of injection. These lesions, however, were only the local cystic areas filled with hemorrhagic, lipid material. There was a minimal chronic reaction in the stroma about some of these cysts.

In this strain of mice spontaneous tumors occur infrequently in the lung and liver, and a few mice develop leukemia. A complete autopsy was made on each mouse injected with the benzene-soluble when this study began. However, there was no suggestion that visceral neoplasms were occurring as a result of this benzene-soluble. Later in the

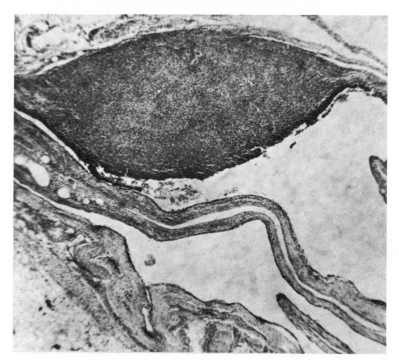

F_IG_. 2. M150–23. Cyst in subcutaneous tissue of a mouse injected 194 days previously with 5.0 mg of benzene-soluble obtained from Texas City. A fibrosarcoma has developed in a local area of the cyst wall. H&E stain, X 25.

experiment the examination of the mice was restricted to a histologic study of the local lesion in the subcutaneous tissue. Mice without a gross lesion, after being observed for a year, were killed and examined macroscopically.

The tumor occurring in the subcutaneous tissue at the site of injection of the benzene-soluble was a fibrosarcoma (Fig. 5). Variations occurred in the cellular structure of these tumors. Cells in mitoses and multinucleated giant cells frequently were present in local areas in some of the tumors. One mouse developed a squamous cell carcinoma. No metastases were present. A few of the mice had a lymphoma or a hemangioma in the liver. Such tumors are spontaneous in this strain of mice.

Four of these fibrosarcomas were transplanted subcutaneously into CFW mice. One grew rapidly in the subcutaneous tissue. An implant 1 to 2 mm was a cm in diameter within 8 to 10 days. This tumor also grew rapidly when transplanted intraperitoneally. The histologic

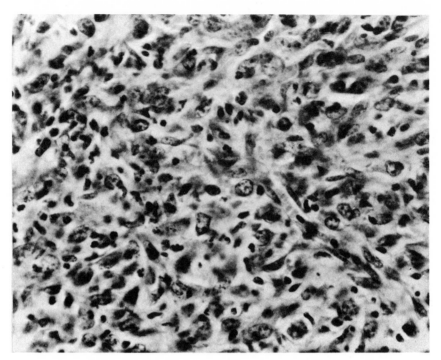

Fig. 3. M150–23. The fibrosarcoma shown in Fig. 2. H&E stain, X 400.

structure of the transplanted tumor was similar to that of the original tumor.

DISCUSSION

A carcinogen is present in the particulate matter obtained from the air in a petrochemical industrial area of Texas City, Texas, and when injected as a benzene-soluble into the subcutaneous tissue of mice produces a fibrosarcoma. Air pollutants obtained from other areas in the state, prepared in a similar manner and injected into the same strain of mice, did not produce tumors.

The particulate matter obtained between 1965 and 1968 rarely produced tumors. However, in the latter part of 1968 and in 1969 a significant number of tumors occurred. There was at this time a corresponding increase in the benzene-soluble and the benzo(a)-pyrene, as shown in Table 1 and 2. The control observations, in which benzo(a)pyrene in varying concentrations was injected subcutaneously. resulted in fewer fibrosarcomas than occurred from the corresponding amount of calculated benzo(a)pyrene that was present in

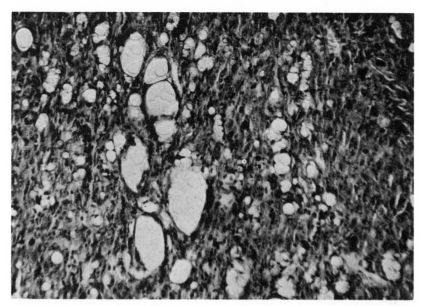

Fig. 4. M139–8. A fibrosarcoma in subcutaneous tissue of a mouse injected 137 days previously with 10.0 mg of benzene-soluble obtained from Texas City. Observe spaces in tumor resulting from the cottonseed oil in which the benzene-soluble was suspended. H&E stain, X 137.

the benzene-solubles. This observation would suggest that carcinogens other than benzo(a)pyrene were present in the benzene-solubles prepared from this air pollutant.

Hueper et al. (1), in their study of the carcinogenic action of benzene-solubles referable to their content of benzo(a)pyrene, also observed a similar variation and stated, "This discrepancy suggests that a definite part of the carcinogenic action of this fraction cannot be attributed to 3–4 benzpyrene but must be due to other carcinogenic agents." Sawicki (4) observed that "usually for each type of pollution, the amounts of polynuclear aromatic hydrocarbons are roughly proportional to the amount of 3,4 benzpyrene—that is, with large concentrations of 3,4 benzypyrene, we find larger concentrations of the other polynuclear aromatic hydrocarbons." No attempt at this time has been made to identify the specific carcinogens, other than benzo(a)pyrene, in the air pollutants we collected.

The percentage of fibrosarcomas observed in the subcutaneous tissue of the mice injected with the benzene-soluble obtained in 1969 is shown in Table 4. The 10.0 mg sample injected in September, 1968, induced 3% tumors, while the 10.0 mg sample injected in October,

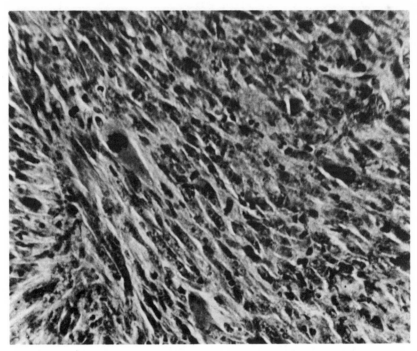

Fɪɢ. 5. M139–8. A portion of the fibrosarcoma shown in Fig. 4 with none of the lipid vacuoles. H&E stain, X 350.

1969, induced 50% tumors. The 3 samples, 2.5, 5.0, and 10.0 mg, injected in October, 1969, were collected at the same time, and their respective percentage of tumors was 40, 61, and 50. The failure of tumors to develop quantitatively in reference to the amount of carcinogen injected is recognized in experimental studies. Although such a factor no doubt is present in this study, additional data are needed to determine whether a specific carcinogen or cocarcinogen was present in the particulate matter collected at one period and not present at another time. Such variations in particulate matter would be expected in an area where there are several chemical plants of various kinds. The carcinogenic action of the benzene-soluble is much higher in our study than that observed by Hueper et al. (1) in the 8 cities which they studied.

Benzo(a)pyrene has been universally found by Sawicki (5,7) in the air of more than 130 urban and nonurban areas in this country. The concentrations varied from 0.01 to 75 μg per 1000 m³ of air, while in this study they varied from 1.4 to 4.0 μ per 1000 m³ of air in 1969. In the airborne particulate the percentage of benzo(a)pyrene varied from 0.00001 to 0.041; and in the benzene-soluble fractions of airborne particulate the percentage varied from 0.00009 to 0.3. The corresponding averages for 1969 in our study are 0.00023 to 0.00034 and 0.04 to 0.0357. Our data are based upon a continuous collection of air samples 6 days out of each week, thus negating the daily variations in air pollutants that is known to occur.

In our study the failure of tumors to develop in mice when lamp black was injected subcutaneously, while tumors did develop when the benzene-soluble prepared from this same lamp black was injected subcutaneously, emphasizes some of the basic problems encountered in air pollutant experiments. Similar observations referable to the occurrence of tumors with carbon black and adsorbed benzo(a)pyrene have been made by Nau et al. (3). They observed that "carbon blacks, channel or furnace, can adsorb and bind known carcinogens so effectively that no tumors are produced when the adsorbed material is injected subcutaneously. . . . The benzene extractives of furnace blacks when injected in oil into mice lead to tumor formation."

The chemical problems encountered in the identification of specific carcinogens in air pollutants are discussed in a recent publication by Sawicki (6). The carcinogenic action of air pollutants in the subcutaneous tissue of mice may be of value in evaluating this problem of carcinogens in urban air. However, additional studies are needed to correlate such observations in the subcutaneous tissue of mice with those in the respiratory tract of man and animal.

REFERENCES

1. Hueper, W. C., P. Kotin, E. C. Tabor, W. W. Payne, H. Falk and E. Sawicki: Carcinogenic bioassays on air pollutants. *Arch. Path.*, 74: 89–116, 1962.

2. Kotin, P., H. L. Falk, P. Mader and M. Thomas: Aromatic hydrocarbons. I. Presence in the Los Angeles atmosphere and the carcinogenicity of atmospheric extracts. *Arch. Indust. Hyg.*, 9: 153–163, 1954.

3. Nau, Carl A., J. Neal and V. A. Stembridge: A study of the physiological effects of carbon black. III. Adsorption and elution potentials; subcutaneous injections. *Arch. Environ. Health*, 1: 512–533, 1960.

4. Sawicki, Eugene: Airborne carcinogens and allied compounds. *Arch. Environ. Health*, 14: 46–53, 1967.

5. ————: Analysis for airborne particulate hydrocarbons. Their relative proportions as affected by different types of pollution. *In*: Symposium, Analysis of Carcinogenic Air Pollutants, National Cancer Institute, Monograph #9, pp. 201–220, August, 1962.

6. ————: Fluorescence analysis in air pollution research. *Talanta*, 16: 1231–1266, 1969.

7. Sawicki, Eugene, W. C. Elbert, T. R. Hauser, F. T. Fox and T. W. Stanley: Benzo(a)pyrene content of the air of American communities. *Amer. Industr. Hyg. Ass. J.*, 21: 443–451, 1960.

8. Sawicki, E., T. W. Stanley, W. C. Elbert and J. D. Pfaff: Application of thin layer chromatography to the analysis of atmosphere pollutants and determination of benzo(a)pyrene. *Anal. Chem.*, 36: 497–502, 1964.

9. Stanley, Thomas, M. J. Morgan and James E. Meeker: Thin layer chromatographic separation and spectrophotometric determination of benzo(a)pyrene in organic extracts of airborne particulates. *Anal. Chem.*, 39: 1327–1329, 1967.

10. Waller, R. E.: The benzpyrene content of town air. *Brit. J, Cancer*, 6: 8–12, 1952.

Microflora of the Respiratory Surface of Rodents Exposed to "Inert" Particulates

Mario C. Battigelli, MD; David A. Fraser, ScD; and Homer Cole

To test the possible favoring effect of inhaled inert dust on the spontaneous microflora of respiratory surfaces of rodents, groups of mice, hamsters, and rats were exposed to graphite (1 mg/cu m) and calcium sulfate (1 mg/cu m) for 12 hours daily, for periods up to four months. A limited number of hamsters have been exposed to high concentrations of graphite dust (10 mg/cu m) continuously (23 hr/day) up to 32 consecutive days. In general, particulate aerosols do not appear to enhance colonization of bacteria or stem bronchi and air spaces. However, a fraction of the hamsters exposed to graphite, 10 mg, developed *Mycoplasma pulmonis*. These observations offer an encouraging hypothesis in support of the possible effect upon *Mycoplasma* infection following exposure to particulate pollutants.

The importance of inert dust in the development of chronic respiratory injury has received renewed emphasis from the standpoint of both industrial[1] and urban air pollution.[2] Yet, the mechanism whereby inert dust at relatively low concentrations may induce pulmonary injury remains obscure. At the levels of contamination commonly present in air of urban and industrial settings, dust aerosols display a variety of behaviors which affect the chemicophysical characteristics of the respirable environment.[3] In this context, an increase in fog frequency, photo-oxidation phenomena, interference with visibility, and even more direct influences on the local climate have been traced to the ubiquitous particulate contamination. The human lung is a most efficient trap of dust[4] which, once deposited, is handled by the cleansing mechanism in a multitude of ways, differing in type and quantity according to the aerosol composition and its intensity for the exposed individual.[5-7]

The actual details of cellular functions and of anatomical pathways for dust handling have not modified

Table 1.—Allocation of Animals

	Control (Ambient Air)	Graphite Dust (1 mg/cu m)	Sulfate Dust (1 mg/cu m)	Graphite Dust (10 mg/cu m)
Mice	15	15	...	...
Rats	108	31	...	...
Hamsters	115	90	74	24

Table 2.—Rats: Frequency of Bacteria Isolates From Stem Bronchi

	Control		Graphite 1 mg/cu m	
Days	N*	%	N*	%
112	13	38	12	33
All†	100	26	28	35

* Number of animals processed.
† Irrespective of duration of treatment.

Table 3.—Rats: Frequency of Bacteria Isolates From Whole Lung Homogenate

	Control		Graphite 1 mg/cu m	
Days	N*	%	N*	%
112	13	23	12	17
All†	71	15	28	14

*† See footnotes for Table 2.

Table 4.—Hamsters: Frequency of Bacteria Isolates From Stem Bronchi

	Control		Graphite 1 mg/cu m		Sulfate 1 mg/cu m	
Days	N*	%	N*	%	N*	%
112	15	27	14	7	15	27
All†	59	12	57	12	59	14

*† See footnotes for Table 2.

Table 5.—Hamsters: Frequency of Bacteria Isolates From Lung Homogenate

	Control		Graphite 1 mg/cu m		Sulfate 1 mg/cu m	
Days	N*	%	N*	%	N*	%
112	15	20	14	0	15	0
All†	59	14	57	11	59	1

*† See footnotes for Table 2.

Table 6.—Hamsters: Frequency of Bacterial Genera Isolated From Whole Lung Homogenates, Irrespective of Duration of Treatment

Treatment Group	No. of Animals Processed	No. of Animals With Positive Culture	Staphylococcus epidermidis	α-Hemolytic Streptococcus	Alcaligenes fecalis	Pseudomonas	Corybacterium	Mycoplasma pulmonis
Control	59	8	3	1	1	2	2	0
Calcium sulfate 1 mg/cu m	59	3	1	0	0	0	0	0
Graphite 1 mg/cu m	57	6	4	2	1	0	2	0
Graphite 10 mg/cu m	24* 9*	6†	0	0	0	0	0	5

* Twenty-four animals processed for bacteriological studies, nine for *M pulmonis* isolation.
† Five animals with *Mycoplasma*. one with bacterial growth.

substantially the classical description formulated by Macklin in 1955,[8] although additional hypothesis have gained the attention of the experts.[9]

The action of inert dust on the airways of volunteers exposed, has been repeatedly studied.[10-12] However, these effects have been mostly limited, in man, to the acute phase of inhalation, while the chronic consequences are still a matter of speculation or indirect epidemiological hints.[13] Certainly the contribution of dust inhaled at these low concentrations to the development of respiratory tract infections has not been verified by experimental evidence.

In view of this it was thought that monitoring of the microflora endogenous to the respiratory surfaces of rodents exposed to dust could provide information on this possible effect.

Materials and Methods

Mice, rats, and hamsters have been exposed to graphite dust of respirable size at concentration of 1 mg/cu m for 12 hours a day up to four months consecutively. In addition, hamsters have been exposed to sulfate dust of respirable size for the same length of time. Finally a group of hamsters have been exposed to graphite dust at 10 mg/cu m for 23 hours a day for 32 consecutive days. The details of the exposure chambers have been given in a previous communication as well as the details on animal processing for the monitoring of surface microflora.[14] The bacteriological studies were obtained from swabates of stem bronchi (rats and hamsters) and from whole right lung homogenate (mice, rats, and hamsters). The inocula were quantitatively plated on blood agar, desoxicholate, and Staphylococcus 110 (Baltimore Biological Laboratories). The lung homogenate was also plated, in a limited number of animals, on Myco-

plasma agar (Robbin Laboratories, Inc).

The number and species of animals processed are summarized in Table 1.

Results

The monitoring of bacterial flora of the bronchi as well as of the lung tissues (whole right lung homogenate) does not present any trend which consistently and significantly differentiated the treated animals from the respiratory control (Tables 2, 3, 4, and 5). An increase in bacterial isolations from the treated animals, in regard to their respective controls, from both bronchi and lungs, is observed occasionally in rats and hamsters, but this trend is not consistent and it is not present at 112 days nor is it reflected in the "total" groups, that is, irrespective of duration of treatment. Mice lung never yields positive culture, either from control or treated animals.

The density of the bacterial population also does not differentiate the treated groups from their controls in any consistent way. Furthermore, the bacterial genera are apparently not affected in any selective and enhancing way, at any of the two sampling sites with one exception (Table 6).

This exception is manifested by the identification of *Mycoplasma pulmonis* in the lung homogenate of hamsters treated with graphite, 10 mg/cu m. In this group five (56%) of the animals examined yielded positive culture of *Mycoplasma*, an infectious agent which is not retrieved from control hamsters and mice, although it is consistently isolated from rats, both control and test animals with equivalent frequency.

Comment

The observation of frequent *Mycoplasma* colonization of lungs of hamsters treated with graphite dust 10 mg/cu m admittedly requires additional verification before it can be accepted as a relevant finding. However, the fact that *Mycoplasma* infection is recognized to be a probable factor of rodent's bronchitis, lends considerable interest to this. It is well known that the chronic nature of this infection in rodents is peculiarly susceptible to being enhanced by concomitant factors of nonspecific injury.[15]

Although, as commented by Clyde, there is no satisfactory animal model reproducing chronic obstructive respiratory disease parallel to the human experience,[16] *Mycoplasma* may be considered an appropriate infectious factor, in rodents, in determining chronic respiratory injuries. If the contributory importance of inert dust is verified in further experiments, a model of injury to test synergism between physical (dust), and biological *(Mycoplasma)* factors of pulmonary injury may herewith be available.

References

1. Lowe CR: Industrial bronchitis. *Brit Med J* 1:463-468, 1969.
2. Lambert PM, Reid DD: Smoking, air pollution and bronchitis. *Lancet* 1:853-857, 1970.
3. Pierrard JM: Environmental appraisal-particulate matter, oxides of sulfur, and sulfuric acid. *J Air Pollut Contr Assoc* 19:632-637, 1969.
4. Druett HA: The inhalation and retention of particles in the human respiratory system, in *Airborne Microbes, 17th Symposium, Society General Microbiology, London, April, 1967.* London, Cambridge University Press, 1967, pp 162-202.
5. Deposition and retention models for internal dosimetry of the human respiratory tract, Task Group on Lung Dynamics. *Health Phys* 12:173-207, 1966.
6. Dyson ED, Beach SA: The movement of inhaled material from the respiratory tract to blood: An analogue investigation of the new lung model. *Health Phys* 15:385-397, 1967.
7. Hatch TS, Gross P: *Pulmonary Deposition and Retention of Inhaled Aerosols.* New York, Academic Press Inc, 1964.
8. Macklin CHC: Pulmonary sumps, dust accumulations, alveolar fluid and lymph vessels. *Acta Anat* 23:1-33, 1955.
9. Casarett LJ, Milley PS: Alveolar reactivity following inhalation of particles. *Health Phys* 10:1003-1010, 1964.

10. Norris RM, Bishop JM: The effect of calcium carbonate dust on ventilation and respiratory gas exchange in normal subjects and in patients with asthma and chronic bronchitis. *Clin Sci* 30:103-115, 1966.
11. DuBois AB, Dautrebande L: Acute effects of breathing inert dust particles and of carbachol aerosol on the mechanical characteristics of the lungs in man: Changes in response after inhaling sympathomimetics aerosols. *J Clin Invest* 37:1746-1755, 1958.
12. McDermott M: Acute respiratory effects of the inhalation of coal dust particles. *J Physiol* 162:53, 1962.
13. Speizer FE: An epidemiological appraisal of the effects of ambient air on health: Particulates and oxides of sulfur. *J Air Pollut Contr Assoc* 19:647-656, 1969.
14. Battigelli MC, Cole HM, Fraser DA, et al: Long-term effects of sulfur dioxides and graphite dust on rats. *Arch Environ Health* 18:602-608, 1969.
15. Edward DGF: The pleuropneumonia group of organisms: A review together with some new observations. *J Gen Microbiol* 10:27-64, 1954.
16. Clyde WA Jr: An experimental model for human *Mycoplasma* disease. *Yale J Biol Med* 40:436-443, 1968.

Suspended Particulate Air Pollution

Relationship to Mortality From Cirrhosis of the Liver

Warren Winkelstein, Jr., MD, and Michael L. Gay, MD

The disease effects of suspended particulate air pollution have been under study in Buffalo and its environs for several years. The community was classified into four pollutant areas by means of measurements obtained from high-volume samplers located at 21 stations and operated over a two-year period. Economic classification was achieved by grouping of census tracts into five levels on the basis of median family income. Age and sex specific death rates among whites for cirrhosis of the liver for the pericensal period, 1959 to 1961, showed a graded positive association with suspended particulate air pollution level when economic status was controlled. The association was strongest in the lowest economic groupings. The findings indicate that urban particulate air pollution may contain toxic agents capable of acting directly or synergistically to cause cirrhosis of the liver.

IN prior papers, we reported results of studies done in Erie County, New York, which demonstrated an association between suspended particulate air pollution levels and mortality from a number of chronic diseases.[1-4] For several, the relationship was strongly positive. These included gastric cancer, prostatic cancer, and selected respiratory diseases. For cerebrovascular and cardiovascular disease, the association was positive but weaker. During the course of these analyses, we also noted differences in mortality from cirrhosis of the liver, International Statistical Classification (ISC)

581, among the areas under study. Although the association of cirrhosis of the liver to alcohol ingestion and certain toxic chemicals has been repeatedly demonstrated, few community-wide studies to investigate its association with other environmental factors have been undertaken. Such studies are important since cirrhosis of the liver is a major cause of disability and death in the United States. Continuing emphasis on studies of alcoholism has retarded the full elucidation of the epidemiology of cirrhosis and almost halted the search for other etiologic factors. The aim of this paper is to describe a previously undemonstrated relationship between cirrhosis mortality and suspended particulate air pollution.

Method

The relationship of suspended particulate air pollution to mortality has been under study for several years in Erie County, New York, an area including the city of Buffalo and its immediate environs. Suspended particulate levels were determined by high-volume samplers located at 21 air sampling stations randomly scattered over the study area. Air samples were obtained during the two-year period from July 1961, through June 1963. Subsequently, isopleths conforming to proposed air quality criteria for New York state were constructed for four pollutant levels. Average annual sex and age specific death rates according to residence at time of death were then computed for the pericensal period, 1959 to 1961.

Since economic status and air pollution levels are frequently inversely associated, an attempt has been made to control for this factor in all analyses of data from the Erie County Air Pollution Study. Thus, the study population was classified into five economic areas on the basis of median family income of each census tract. The cutoff points were determined by a method which purports to optimize the homogeneity of the economic groupings.[3] This permitted the comparison of mortality rates between areas of essentially the same economic status, but differing air pollution. In the present study, nonwhites were excluded since their numbers in the study area were too small to produce meaningful rates.

Results

Mortality rates for cirrhosis of the liver among men and women within the two age groups studied are higher than comparable rates for the total United States as shown in Table 1. They are, however, similar to rates in other major metropolitan areas.[5] The male rates in both the study area and the United States were two to three times as high as the female rates. The differences between economic levels shown in Tables 2 and 3 are, likewise, consistent in direction and degree with the findings of other investigators.[6,7]

Since the mortality rates for cirrhosis in both men and women 70 years and older were only 23% and 27% higher, respectively, than the rates among those 50 to 69 years old, and since the numbers of deaths available for analysis among older men and both female age groups were relatively small, we elected to pool the two age groups. Furthermore, as shown in Table 4, the proportions of older persons in the various subgroups are fairly constant, and their distribution is such that a bias due to age would not occur as a result of pooling.

The association between cirrhosis mortality, economic level, and air pollution for white males is shown in Table 2. The marginal totals reveal a strong inverse association between economic level and cirrhosis mortality and a similar strong, but positive, association between cirrhosis mortality and suspended particulate air pollution. When the association of economic level and cirrhosis is examined within each air pollution level, the trend is maintained except in the lowest pollution area. When the association of air pollution level and cirrhosis is examined within each economic level, the positive trend is most strongly evident in the two

Table 1.—Average Annual Death Rates per 100,000 Population From Cirrhosis (ISC 581)*

Age (yrs)	Men				Women			
	Buffalo		United States		Buffalo		United States	
	No.	Rate	Rate		No.	Rate	Rate	
50-69	169	75	50		55	22	19	
≥70	55	92	52		23	28	22	
Total	224	79	50		78	23	20	

* According to selected ages and sex, whites only, Buffalo and environs, 1959 to 1961, and the United States, 1960.

lower economic areas although it is preserved within each of the five strata. In economic level 2, (median family income of $5,175 to $6,004) with a total white male population over 50 years of 30,550 and population representation in each of the four air pollution levels, the rates show a graded increase from 28 per 100,000 in the lowest air pollution level to 173 per 100,000 in the highest.

The data for women is presented in Table 3. The inverse association between cirrhosis mortality and economic level is less pronounced than for men. In fact, when the association is examined within each air pollution level, it is clearly apparent only in air pollution level 3. On the other hand, the positive association between cirrhosis mortality and suspended particulate air pollution remains strong in the three lower economic levels, although it disappears in the two upper. In economic level 2, with a white female population of 35,908, the association is graded with an increase from 11 per 100,000 in the lowest pollution area to 47 per 100,000 in the highest.

Since the International Classification of Diseases distinguishes between cirrhosis deaths with and without mention of concurrent alcoholism, we have examined our data to see whether the patterned associations differ when the analysis is carried out for the separate rubrics.[8] Forty-four percent of the cirrhosis deaths in this study were reported with concurrent alcoholism. The patterns of association with both air pollution and economic level were closely similar for the two groups, and in Table 5 the death rates for all men and women 50 years and older in economic level 2 are shown for the four air pollution levels represented.

Comment

In the preceding reports from the Erie County Air Pollution Study, we indicated some of the major problems involved in interpreting the findings.[1-4] These include the indirect nature of the associations; the inability to take into account such personal characteristics of the study population as smoking habits, occupation, and in this analysis, alcohol consumption; and the possibility that the patterned associations could be due to selective migration. Nevertheless, the strength of the associations, their internal consistency, their biologic plausibility, and their comparability with other research findings lend support to their validity.

In economic level 2, wherein there is population representation in all four air pollution levels, and which has a relatively narrow range of median family income for its census tracts ($829), both men and women show a regularly graded association with cirrhosis mortality rates in the highest pollution area more than six times that in the lowest for men and more than four times that in the lowest for women. The consistency of the pattern of association between men and women also indicates that occupational exposure to toxic agents is not the major determinant of the pattern.

It is, perhaps, worth noting that the ratio of the male to female death rates rises from 1.7 in the lowest polluted area to 3.9 in the highest. In economic level 2,

Table 2.—Average Annual Death Rates per 100,000 Population From Cirrhosis of the Liver*

| Economic Level | Air Pollution Level | | | | |
	1 (Low)	2	3	4 (High)	Total
1 (low)	...	129 (3)†	129 (28)	359 (28)	185 (59)
2	28 (4)	82 (32)	144 (38)	173 (21)	103 (95)
3	...	48 (14)	87 (13)	81 (4)	63 (31)
4	37 (9)	33 (10)	83 (8)	...	43 (27)
5 (high)	26 (6)	17 (4)	95 (2)	...	25 (12)
Total	31 (19)	51 (63)	119 (89)	213 (53)	79 (224)

* According to economic and air pollution levels, white men 50 years and older, Buffalo and environs, 1959 to 1961.
† Numbers in parentheses indicate deaths.

Table 3.—Average Annual Death Rates per 100,000 Population From Cirrhosis*

| Economic Level | Air Pollution Level | | | | |
	1 (Low)	2	3	4 (High)	Total
1 (low)	...	0 (0)†	37 (8)	69 (4)	39 (12)
2	11 (2)	19 (9)	30 (9)	47 (6)	24 (26)
3	...	17 (6)	25 (4)	57 (3)	23 (13)
4	31 (8)	21 (8)	27 (3)	...	25 (19)
5 (high)	11 (3)	17 (5)	0 (0)	...	13 (8)
Total	18 (13)	18 (28)	29 (24)	55 (13)	23 (78)

* According to economic and air pollution levels, white women 50 years and older, Buffalo and environs, 1959 to 1961.
† Numbers in parentheses indicate deaths.

Table 4.—Ratio of Population 70 and Older to Population 50 to 69*

| Economic Level† | Air Pollution Level‡ | | | | | | | |
| | 1 (Low) | | 2 | | 3 | | 4 (High) | |
	M	F	M	F	M	F	M	F
1 (low)	...	...	0.46	0.60	0.37	0.43	0.33	0.42
2	0.30	0.41	0.33	0.36	0.26	0.26	0.26	0.27
3	...	...	0.25	0.28	0.24	0.29	0.25	0.28
4	0.20	0.27	0.24	0.36	0.21	0.28	...	...
5 (high)	0.21	0.32	0.22	0.33	0.22	0.22	...	...

* According to sex, economic level, and suspended particulate air pollution level.
† Based on median family income for each census tract: 1, $3,005-$5,007; 2, $5,175-$6,004; 3, $6,013-$6,614; 4, $6,618-$7,347; and 5, $7,431-$11,792.
‡ Based on two year average suspended particulate levels: 1, <80µg/m³/24hr; 2, 80µg-100µg/m³/24 hr; 3, 100µg-135µg/m³/24 hr; and 4, >135µg/m³/24 hr.

Table 5.—Average Annual Death Rate per 100,000 Population From Cirrhosis*

Air Pollution Level	Cirrhosis Without Mention of Alcoholism	Cirrhosis With Mention of Alcoholism
1 (low)	9 (3)†	9 (3)
2	32 (27)	16 (14)
3	48 (27)	36 (20)
4 (high)	64 (16)	44 (11)
Total	37 (73)	24 (48)

*According to suspended particulate air pollution level, presence or absence of concurrent alcoholism; men and women 50 and over, Buffalo and environs, 1959-1961, economic level 2 only.
† Numbers in parenthesis indicate deaths.

99

the corresponding ratios are 2.5:3.7. This suggests that air pollution acts synergistically with other agents to produce cirrhosis mortality and that men may have greater exposure to hepatotoxic agents, occupationally or through alcohol consumption, than women. This would be fully consistent with numerous original studies of hepatotoxic agents.[9-11] On the other hand, a number of established hepatotoxic chemicals are known to be components of urban suspended particulate air pollution.[12] In fact, both primary liver cancer and cirrhosis have been experimentally induced in animals by injection of extracts of particulate pollution collected in Cincinnati and other cities.[13,14] There were insufficient deaths from primary cancer of the liver during the three-year study period to permit an analysis of their association with air pollution in this study.

Mortality from cirrhosis of the liver in Buffalo was investigated some years ago by Lilienfeld.[6] His analyses revealed a similar socioeconomic gradient and sex differential to that reported here. He interpreted his observations as indicative that occupational exposures to toxic materials might be major factors in cirrhosis mortality. His conclusions were subsequently questioned. However, we believe that Lilienfeld's conclusions, as well as those of others who hold alcohol consumption responsible for the bulk of cirrhosis mortality, are consistent with the hypothesis that air-borne toxic agents act directly to produce cirrhosis as well as synergistically to augment the effects of other hepatotoxins.

The findings of this study suggest that liver disease should not be ignored in experimental or epidemiological research on the effects of air pollution. While it would be presumptuous to assert that the results reported here are sufficient to establish a causal relationship between exposure to urban particulate air pollution and cirrhosis of the liver, they appear to us to warrant early efforts at replication wherever the necessary data exist.

This study was supported in part by research grant 1 PO1 CA11428-01 from the National Cancer Institute and AP-00181 from the Division of Air Pollution, Public Health Service.

References

1. Winkelstein W Jr, Kantor S: Stomach cancer: Positive association with suspended particulate air pollution. *Arch Environ Health* 18:545-547, 1969.

2. Winkelstein W Jr, Kantor S: Prostatic cancer: Relationship to suspended particulate air pollution. *Amer J Public Health* 59:1134-1138, 1969.

3. Winkelstein W Jr, Kantor S, Davis WD, et al: The relationship of air pollution and economic status to total mortality and selected respiratory system mortality in men: I. Suspended particulates. *Arch Environ Health* 14:162-171, 1967.

4. Winkelstein W Jr, Gay ML: Arteriosclerotic heart disease and cerebrovascular disease: Further observations on the relationship of suspended particulate air pollution and mortality in the Erie County Air Pollution Study. *Proc Inst Environ Studies* 16:441-447, 1970.

5. Kramer K, Kuller L, Fisher R: The increasing mortality attributed to cirrhosis and fatty liver in Baltimore (1957-1966). *Ann Intern Med* 69:273-282, 1968.

6. Lilienfeld AM, Korns RF: Some epidemiologic aspects of cirrhosis of the liver: A study of mortality statistics. *Amer J Hyg* 52:65-81, 1958.

7. Terris M: Epidemiology of cirrhosis of the liver: National mortality data. *Amer J Public Health* 57:2076-2088, 1967.

8. *Manual of the International Statistical Classification of Diseases, Injuries, and Causes of Death (International Classification of Diseases)*. Geneva, World Health Organization, 1957.

9. Moon VH: Experimental cirrhosis in relation to human cirrhosis. *Arch Path* 18:381-424, 1934.

10. Klatskin G: Effect of alcohol on the liver. *JAMA* 170:1671-1676, 1959.

11. Galambos JT: Alcohol and liver disease. *Amer J Dig Dis* 14:477-490, 1969.

12. *US Air Quality Criteria for Particulate Matter*, publication AP-49. US Dept of Health, Education, and Welfare, Public Health Service, Environmental Health Service, National Air Pollution Control Administration, 1969.

13. Epstein SS, et al: The carcinogenicity of organic particulate pollutants in urban air after administration of trace quantities to neonatal mice. *Nature* 212:1305-1307, 1966.

14. Epstein SS, Joshi S, Andrew J, et al: The carcinogenicity of organic particulate pollutants in urban air after administration of trace quantities to neonatal mice. *Nature* 212:1305-1307, 1966.

Lysosomes and the Toxicity of Particulate Pollutants

A. C. Allison, MD,

The pathogenesis resulting from inhaled particulate pollutants centers around their effects upon the macrophage. An analysis of pathogenesis must proceed in two stages: first, determining how particles kill macrophages, and second, determining how this relates to fibrogenesis. Silica (silicon dioxide) alters the secondary lysosomal membrane, making it permeable to the contained lysosomal enzymes. Silica surfaces alter the lysosomal membrane through hydrogen bonding. Asbestos produces similar effects but not to the extent shown by silica. Fibrogenesis occurs through macrophage damage and release of as yet unknown factors.

Many particles inhaled into the lung are more or less innocuous, eg, the carbon particles that remain in macrophages for long periods. However, some inhaled particles, especially silica and asbestos, stimulate a severe fibrogenic reaction, and there are indications that these and other particles also represent an important portal of introduction of carcinogens such as benzo[a]pyrene into the pulmonary tissues. I have been concerned with silica and asbestos, and wish to describe experiments on the cytotoxic effects of these particles. It is necessary, however to begin with a brief description of the materials themselves.

Silicon and oxygen are the commonest elements by weight in the lithosphere; they can exist as silica (silicon dioxide) or in combination with other elements in the form of silicates such as asbestos, mica, kaolin, or talc. Silica is composed of tetrahedra in which a silicon atom lies centrally and four oxygen atoms are each shared with a neighboring atom of silicon. Differences in the spatial relationships of the tetrahedra, as shown by the x-ray diffraction pattern, account for the various physical forms of silica. If the pattern is irregular, the silica is termed amorphous, as in the case of quartz glass (vitreous silica), obtained by melting; and diatomaceous earth (kieselguhr). Each type of crystalline silica has a characteristic regular array of tetrahedra in three dimensions. Quartz is the stable crystal at atmospheric pressure, whereas two other crystal types, tridymite and cristobalite, formed at higher temperatures than quartz, are metastable. Two further types of crystalline silica, coesite and stishovite, are formed at very high temperatures and pressures. Both have been synthesized[1] and isolated as natural minerals from Meteor Crater, Ariz. Stishovite is unique in not having the usual tetrahedral form; it has, instead, an octahedral structure in which six shared oxygen atoms are attached to each atom of silicon, a disposition found also in the rutile variety of titanium dioxide.[3] Stishovite has the highest density, then coesite, then the low pressure forms, with tridymite and cristobalite having more open structures than quartz.

The name "asbestos" is applied to minerals that break down into fibers when crushed or processed. This property results from a molecular structure in which the units are linked together most strongly in one direction. A number of fibrous mineral silicates differing in chemical composition show the properties of asbestos. These include two main mineral groups: (1) pyroxenes–chrysotile, and (2) amphiboles–crocidolite, amosite, anthophyllite, tremolite, and actinolite. Chrysotile, or white asbestos, is the most common type, making up more than 90% of that used commercially. Extensive deposits are mined in many parts of the world, especially Canada, Russia, and Rhodesia. Chrysotile is a hydrated magnesium silicate in the form of scrolls or tubes which vary in diameter from 150 to 400 angstroms (Fig 1). The mineral forming these tubes is crystalline; that in the hollow centers and between the tubes is amorphous. The tubes do not adhere together very strongly, and this accounts for the ease with which chrysotile can be broken down into very fine fibers. The tubes are formed from two layers, the inner a continuous network of hexagonally arranged silica units, which is interlocked with the second layer, a brucite-like sheet of octagonal magnesium hydroxide. The spacing of the silica sheet is tighter than required for a perfect fit with the brucite layer, and this accounts for the contortion into tubes or scrolls with the magnesium hydroxide layer on the outside.

The other commercially important asbestos minerals are members of the amphibole group; the molecular structure is essentially the same, with individual types characterized by variations in the composition of the cation chains. The smallest fibers (usually between 800 and 1,000 A in diameter) are solid and coarser than the finest fibers of chrysotile. The structure of the amphiboles is well established.[4] They consist of octahedral cation oxide units arranged in narrow strips and sandwiched between corresponding strips of hexagonally arranged silica tetrahedra. Since the strips are loosely bonded to each other along the edges and faces, fibrous cleavage readily occurs. Of the amphiboles, crocidolite (blue asbestos) is also commercially important. It is sodium ferroso-ferric silicate, mined in South Africa and other countries, and was the first form shown to be carcinogenic. It is very resistant to acid. Another amphibole is amosite, ferrous magnesium silicate, in the form of long brownish-yellow fibers mined in South Africa and used in asbestos cement and heat-insulating products. A third amphibole used commercially on a small scale is anthophyllite, fibrous magnesium silicate. Two other varieties of amphibole asbestos—tremolite (calcium magnesium silicate) and actinolite (calcium magnesium iron silicate)—have less commercial use.

Study of the biological properties of asbestos is complicated by several factors, including the presence of many varieties, the variable contamination of the basic minerals by other materials which may be biologically important, and the fact that biological effects are markedly dependent on fiber size. The longest asbestos fibers normally encountered may be 2 inches or more in length, but the attrition that is unavoidable in separating asbestos fiber from nonfibrous rock results in the presence of fibers so short that the length is hardly greater than the diameter. The finest fibers are always present to a considerable extent in asbestos dust, and these are too small to be resolved in an optical microscope.

Particle size and shape is important in relation to likelihood of inhalation into the lungs, site of deposition, and probability of clearance. This is a complex problem beyond the terms of reference of this article, but it will be obvious that smaller particles have a greater chance of remaining suspended in respired air, penetrating to the alveolar ducts, and so being retained in the lungs, than larger particles. Shape can also be important; highly asymmetric asbestos particles, especially when they are curved (as in the case of chrysotile) are less likely to penetrate deep into the lung than nearly isometric particles, such as silica, of comparable weight. On the other hand, the asymmetry may hinder phagocytosis, and this may account for the fact that asbestos particles reach the pleural mesothelium; this seldom, if ever, happens with silica particles, which are engulfed by alveolar macrophages.

Effects of Toxic Particles on Macrophages

It is generally accepted that the initial event in silicosis is the phagocytosis of silica particles by alveolar macrophages and consequent death of the cells. The particles so released are taken up by other macrophages which are, in turn, killed. In this way death of macrophages continues and stimulates collagen synthesis by fibroblasts in the neighborhood.

Analysis of pathogenesis must therefore proceed in two stages: first determining how silica particles kill macrophages, and, second, determining how this is related to fibrogenesis.

As Marks[5] showed, the cytotoxic effects of silica can be conveniently reproduced in cultures of peritoneal or alveolar macrophages, and the relative toxicity of different forms of silica, and of different dusts, on cell cultures agrees with the pathogenicity and fibrogenic activities of the dusts in vivo.[6,7] When my colleagues (T. Nash, J. Harrington, and M. Birbeck) and I began working on this problem in 1964, there was no satisfactory explanation of silica toxicity. King, is his well-known "solubility" theory, suggested that silicic acid liberated into the tissues from silica particles brings about deposition of collagen. Later observations did not support this interpretation, as King[8] himself pointed out. Curran and Rowsell[9] showed that silica particles implanted into the peritoneum in diffusion chambers do not induce any fibrogenic reaction, although silicic acid is liberated from the chambers. Vigliani and Pernis[10] formulated an autoimmune theory of silicosis, but several workers were unable to obtain experimental evidence in support of this interpretation.[11]

We, therefore, made a detailed study of the effects of toxic and nontoxic particles on cultures of macrophages, using time-lapse phase-contrast cinephoto-micrography, histochemistry, and electron microscopy.[12] Particles of silica, diamond dust, and other materials were rapidly included in phagosomes surrounded by single membranes. Lysosomes became attached to the phagosomes and discharged their enzymes into the phagosomes. So far, there was no difference between the toxic and nontoxic particles. After about 18 hours' incubation, however, clear differences were apparent. The nontoxic particles and associated enzymes were still enclosed in secondary lysosomes, whereas many of the toxic particles and associated lysosomal enzymes had escaped into the cytoplasm. The macrophages that had ingested nontoxic particles were fully extended and moving about freely, whereas many of those exposed to toxic particles were rounded and immobile. Thus, it was evident that silica particles, unlike nontoxic particles, can react with lysosomal membranes and make them permeable. Histochemical studies showing release of peroxidase taken up into lysosomes of macrophages after treatment with silica, confirming our observations with acid phosphatase, have been published by Nadler and Goldfischer.[13]

Conventional biochemical studies of the redistribution of lysosomal enzymes from the particulate to the supernatant fraction could not be carried out because homogenization of normal macrophages in the presence of silica particles showed disruption of lysosomes. Hence, an indirect measure of enzyme release which follows such redistribution after phagocytosis of bacteria[14] was used. We found that several lysosomal enzymes were released into the medium after exposure of macrophages to silica, whereas this was not seen in control cells incubated with nontoxic dusts or silica coated with a protective material, poly-2-vinyl-pyridine-N-oxide (PVPNO). The release of lysosomal enzymes into the medium preceded release of the cytoplasmic enzyme lactic dehydrogenase and cell death as determined by several criteria (trypan blue staining, tetrazolium reduction, and fluorescein ester cleavage). Thus, the lysosomal enzyme release appears to precede rather than follow cell death.

Our observations have been confirmed repeatedly. Silica particles taken up in culture[15,16] or following inhalation exposure[17,18] are initially in membrane-bound vacuoles or secondary lysosomes but are later observed free in the cytoplasm; the latter coincides with morphological signs of damage to mitochondria and other cell organelles. Ingested coal[19] or kaolin[20] produced only minimal cytological changes. The effects of inhaled silica on macrophages progress more slowly than in culture, in which the number of particles or aggregates per cell is higher, but the sequence of events is essentially the same.[18] Release of acid phosphatase and other lysosomal enzymes from macrophages exposed to silica into the medium, and its prevention by PVPNO, has been confirmed by Comolli,[21] Sakabe and Koshi[22] and Beck.[16]

Interaction of Silica
With Biological Membranes

The observations which have been described focus attention on the capacity of silica particles, in contrast to those of many other particles, to react with lysosomal membranes and make them permeable. Silica particles can, in fact, interact with a variety of biological membrane systems. The simplest demonstration is provided by mixing washed erythrocytes with suspensions of silica particles or with silicic acid preparations.[23,24] The erythrocytes are quite rapidly lysed by several fibrogenic forms of crystalline silica, whereas one form which is not fibrogenic (stishovite), and several other types of nonfibrogenic dust of comparable size and surface area, produce very little hemolysis. We have presented reasons[24] for believing that the toxicity of silica is due to the fact that the particles are easily ingested and by interaction with water form on their surface silicic acid which can act as a powerful hydrogen bonding agent.

There are two classes of hydrogen bonding compounds. The larger class comprises hydrogen acceptors such as ethers and ketones with active lone-pair electrons on oxygen or nitrogen. The smaller class comprises hydrogen donors of which only the phenols are important among organic compounds and silicic and some other very weak acids among inorganic compounds. Compounds of one class interact with those of the other, so that it is not surprising that in one class (the former), the members are compatible with living cells, sometimes in very high concentration, whereas those of the other class are damaging.

Model experiments showed that hydrogen bonding of phenolic hydroxyl groups, of the type present in silicic acid, occurs with secondary amide groups of proteins, and this can lead to protein denaturation. However, the interaction with phospholipids is stronger, and we have presented evidence that this is more important in interactions with biological membrane systems.

In view of the interaction with membranes and lysis of erythrocytes, the question arises whether silica particles can kill macrophages or other target cells by interaction with their outer membranes. If relatively large amounts of silica ($>100\mu g/ml$) are added to macrophages in a saline medium lacking protein, such rapid killing by cell membrane lysis is observed. Oxygen consumption falls rapidly, within one hour there is considerable leakage of lactate dehydrogenase and other cytoplasmic proteins into the medium, and within two or three hours most of the cells are dead, as shown by the lack of fluorescein diacetate hydrolysis using the method of Rotman and Papermaster[25] and other criteria. However, this is a completely unphysiological situation, and if silica is added to cells in a medium containing serum protein or bronchial washings, this type of surface interaction does not occur. The silica particles are coated with serum protein, even after repeated washing, as can be shown by agglutination with antibody of appropriate specificity staining with fluorescein-labeled antibody. The particles coated in this way are taken up into lysosomes, and it is only when this coating is digested away after some hours (as shown by the disappearance of staining around the

particles by fluorescein-labeled antibody) that the underlying silica surface is exposed to interact with the lysosomal membrane.

Since most forms of silicic acid in solution are less damaging than silica particles, even though the former have hydrogen-bonding groups, it seems that the rigid structure of the quartz, with many hydrogen-bonding groups arranged in a regular and immovable order on its surface, must be important in damaging cells. Presumably, the formation of multiple bonds distorts the membrane structure and so leads to the breakdown of the membrane.

Prevention of Silica Toxicity by Polyvinylpyridine-N-Oxide (PVPNO)

Evidence in support of the interpretation that hydrogen bonding is important in silica toxicity comes from experiments with PVPNO. It was found by Schlipköter et al[26] that this polymer markedly reduces the amount of fibrous tissue formed after intravenous injection of silica. Other experiments on the inhibition of quartz fibrosis by PVPNO were summarized by Schlipköter[27]; although silica and PVPNO did not have to be given together, they had to be administered in such a way that they reached the same sites. The protective action of PVPNO in animals has been confirmed by numerous other workers and the compound has also been shown to protect cells in vitro against the damaging effects of silica particles.[12,28] Comparisons between protection of cell cultures in vitro and of whole animals in vivo against silica toxicity of fibrogenicity by various polymers[29] led to the conclusion that compounds ineffective in tests on cultured cells would also be ineffective in animals, whereas some polymers which protect macrophages against quartz in cultures are ineffective in animals. This is understandable if some compounds fail to reach or persist in secondary lysosomes of cells in which silica is taken up in vivo, because the compounds are rapidly metabolized or for some reason other than rapid metabolism.

Allison et al[12] reported that PVPNO is taken up by pinocytosis and concentrated in secondary lysosomes in the same way as dextran, povidone (polyvinylpyrrolidone), and other polymers.[30] The uptake of PVPNO into lysosomes of macrophages and other cells has been confirmed by Grundmann,[31] Bruch,[17] Beck and Boje,[32] Bairati and Castano,[33] and others; these observations have been made by histochemistry, electron microscopy, and cell fractionation and autoradiography using labeled polymer. Nash et al[24] pointed out that PVPNO has oxygen atoms which can readily form hydrogen bonds with phenolic hydroxyl groups, so that the polymer can preferentially interact with silicic acid on the surface of silica particles before the latter can attack lysosomal membranes. In support of this interpretation, Dehnen and Fetzer[34] reported that PVPNO protected rat liver lysosomes from damage by quartz.

In 1966 Nash and I synthesized another compound, polyvinylpyridinioacetic acid, which has even greater hydrogen bonding capacity than PVPNO and very efficiently protects macrophages from silica toxicity. This protection was confirmed by Sakabe and Koshi,[22] who also showed that it prevents the increase in acid phosphatase in macrophages following exposure to silica. Holt et al[35] also reported that some poly(2-vinylpyridinium) salts prevented quartz toxicity in cultures. Thus the presence of hydrogen-accepting groups (either N-oxides or pyridinium groups) is frequently associated with protection, although other factors may also be involved. Thus, Holt et al[35] found a difference between the protective effect of syndyotactic and isotactic PVPNO, polymers that differ only in the spatial arrangement of the structural units.

These two facts are sufficient to explain why silica is so toxic to macrophages: the particles are taken up into lysosomes and readily damage lysosomal membranes through hydrogen-bonding interactions. Various secondary reactions may occur. Thus, Munder et al[36] have found a considerable increase in the concentration of lysolecithin, as compared with lecithin, in macrophages damaged by quartz. This could follow activation of the enzyme phospholipase A, which catalyses the reaction lecithin $\rightarrow$ lysolecithin and which is known to be lysosomal. However, the fact that silica lyses erythrocytes (membranes of which do not contain demonstrable amounts of phospholipase A) suggests that this process is unnecessary for interaction of silica with membrane systems, although the formation of surface-active lyso-lecithin could well accelerate damage induced by silica in macrophages.

Suspensions of silica particles readily release enzymes from isolated liver lysosomes in vitro, as Dehnen and Fetzer's[34] experiments and our own have shown; the relatively low temperature coefficient for this release suggests that physico-chemical rather than enzymic reactions are involved.

The second question remains: how is macrophage damage related to fibrogenesis. An interesting lead has been obtained by Heppleston and Styles.[37] They found that macrophages incubated in culture with silica particles released into the supernatant fraction a factor which, when added to fibroblast cultures, stimulated collagen formation as judged by synthesis of hydroxyproline. This stimulation appeared to be due to a specific product of the macrophage-silica interaction. It was not seen in normal macrophages or in macrophages exposed to nontoxic particles (titanium dioxide, ie, anatase) or to silica in the presence of sufficient PVPNO. The nature of the stimulating factor is still unknown, but it seems likely that no direct interaction of particulate silicate with fibroblasts is involved.

Potentiation by Silica of Tuberculous Infections

All published studies of silicotic human subjects have shown a much higher incidence of tuberculosis than in nonsilicotic subjects from the same areas.[38] Before drugs were introduced, tuberculosis was reported as the main cause of death in classical silicosis.[39] The response to chemotherapy of tuberculous infections in patients with pneumoconiosis is also less satisfactory than in subjects without the latter.[40] Moreover, since the work of Cesa-Bianchi[41] it has been recognized that infections with *Mycobacterium tuberculosis* in experimental animals are aggravated by injections or inhalations of silica dust. Such potentiation has been observed with a substrain of BCG in guinea-pigs and with a virulent strain of *M tuberculosis* in rats[38,42] and also with *M kansasii*, a human pathogen, in guinea-pigs.[43] The potentiation by silica is demonstrable in several ways, including establishment of infections with fewer organisms than are required to produce lesions in the absence of silica particles, higher counts of recovered organisms, spread from subcutaneous inoculation sites to the lungs, and fatal outcome of infection only in animals exposed to silica dust.

These results show that in both human subjects and experimental animals that growth and spread of tubercle bacilli are increased by concomitant exposure to silica dust. There is no evidence that silica interferes with immune responses to *M tuberculosis*. Indeed it has been found that animals exposed to silica or silicate actually have greater humoral and cell-mediated immune responses than in the absence of silica or silicate.[44,45] Vaccination with BCG was ineffective in preventing the increased growth of tubercle bacilli in silicotic guinea-pigs,[42] and allergic reactions to tuberculin are greater in

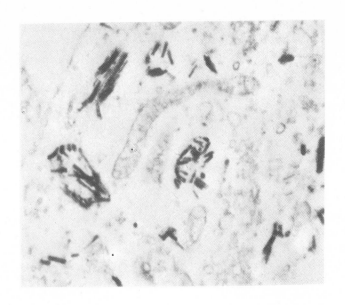

Fig 1.—Electron micrograph of a macrophage that has ingested short fibers of chrysotile (× 20,000). The asbestos fibers are seen in secondary lysosomes.

infected silicotic animals than in infected controls.[38,42,43]

Hence, it seems that some factor other than conventional immunity must be involved in the potentiation by silica of the growth of tubercle bacilli. One possibility is that the response of macrophages to intracellular *M tuberculosis* is altered in the presence of silica. The role of macrophages in protection against tuberculous and other intracellular infections has been extensively discussed by Mackaness.[16] "Activation" of macrophages by infection can increase resistance against apparently unrelated organisms, which helps to illustrate the key role of macrophages in overcoming such infections. Since both intracellular infective organisms and silica particles are present in secondary lysosomes of macrophages, it seemed that the latter might potentiate the growth of tubercle bacilli. To test this possibility Hart and I examined the effect of sublethal doses of silica on the multiplication of *M tuberculosis* in macrophages from normal mice cultured in the absence of immune serum. In such a closed system general effects on host resistance would be excluded and specific effects on macrophages revealed.

We found[17] that growth of a relatively virulent strain of *M tuberculosis* (H37RV) in cultures of macrophages was potentiated by the addition of sublethal doses of silica. Macrophages treated with silica showed earlier multiplication of organisms, and the cells were released

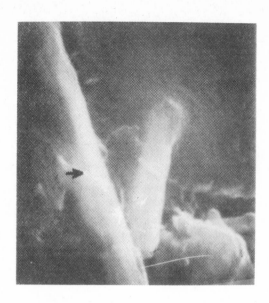

Fig 2.—Stereoscan electron micrograph of a macrophage that has incompletely ingested asbestos fibers (× 50,000). The plasma membrane is reflected at-the point marked by the arrow. The parts of the fibers below this point are enveloped by invaginated plasma membrane and cytoplasm; the upper parts of the fibers are free.

from their glass support into the medium sooner than control macrophages, both infected without silica or treated with silica alone. Mycobacteria are released mainly by death and dissolution of the macrophages in which they grow. There is little doubt that if these two factors—faster multiplication and earlier release from host cells—operate also in vivo they would greatly facilitate the growth and spread of organisms in the presence of silica particles. This may be sufficient to account for the well-documented aggravation of tuberculous infections in both man and experimental animals by exposure to silica dust.

Effects of Asbestos Particles on Cells

Exposure of humans and experimental animals to asbestos has three main consequences: (1) the development of asbestos bodies, in which the typical particles are found, surrounded by iron-containing pigment which on electron microscopy is seen to contain ferritin and hemosiderin; (2) the fibrogenic response characteristic of asbestos (this may be marked, although it is less severe than that produced by comparable amounts of silica); and (3) development of tumors in the lung. These tumors include a high proportion of mesotheliomas arising from the layer

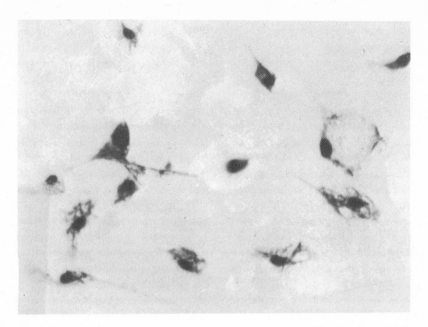

Fig. 3.—Macrophage culture two days after exposure to chrysotile asbestos, showing vacuolation of some cells and dense-staining cytoplasm of others.

of cells lining the pleura (or occasionally the peritoneum)—a type of tumor that is extremely rare in individuals not exposed to asbestos.

Although the presence of many mineral forms of asbestos, variety of sizes, and variability of results depending on experimental conditions makes it difficult to draw general conclusions about effects of asbestos on cells, some overall points are beginning to emerge. The first is that certain types of asbestos can certainly interact with cell membranes. This is shown by the hemolytic effects of asbestos described by Macnab and Harington[48] and Secchi and Rezzonico.[49] Both groups of authors found chrysotile is at least as hemolytic as silica, but Macnab and Har-

ington reported that PVPNO and aluminum oxide, which markedly reduced hemolysis by silica, had less effect on chrysotile hemolysis. Ethylenediaminetetraacetate or simple phosphates markedly reduced chrysotile hemolysis, which is not surprising since the crystal structure of chrysotile is such that the surface consists of magnesium hydroxide. Chrysotile was found by Secchi and Rezzonico to adsorb the erythrocyte membrane enzyme acetylcholinesterase strongly but the cytoplasmic enzyme lactate dehydrogenase only weakly. They suggested that the highly hemolytic activity of chrysotile is related to the adsorptive capacity of the dust for erythrocyte membrane components. Crocidolite, amosite, and anthophyllite were only weakly lytic.

There are apparent discrepancies between descriptions by various authors of effects of asbestos on macrophages and other cells in culture. On the one hand, there is a report by Parazzi et al[50] that crocidolite and chrysotile are very toxic to guinea pig macrophages, crocidolite being more toxic than chrysotile or tridymite dust used for comparison; fibers were more damaging than particles. On the other hand, the cellular changes found by Beck et al[15] and Davis[31] on exposure of organ cultures to chrysotile were relatively slight.

My co-workers and I have examined the effects of various asbestos dusts (International Union Against Cancer [IUCC] standard samples) on cultures of mouse peritoneal macrophages, prepared as previously described,[12] and on organ cultures of mouse peritoneum in the same medium. The peritoneum is lined by mesothelial cells which are involved in asbestos malignancies, so that their reaction to asbestos was worth studying. This preparation was of interest for several reasons—fibroblasts quickly grew out and were found to take up and be affected by asbestos particles much less than mesothelial cells. The first experiment concerned the capacity of cells to take up asbestos fibers of different lengths. Fractions containing high and low percentages of long fibers were added to cells in medium 199 containing 10% fetal calf serum. Independently of the asbestos type, short fibers (<5µ) were readily and completely taken up by phagocytosis,

whereas long fibers (>30µ) were not. The cells were closely attached to or enveloped the ends of the latter, as shown by ordinary and stereoscan electron micrographs (Fig 2), but part of the asbestos fibers remained outside the cell. This was confirmed by phase-contrast cinephotomicrographs. With very long fibers, two or more cells could be seen closely attached to a single fiber, sometimes with apparent continuity of cytoplasm; such attachment may favor syncytium formation. Cells attached to long fibers in the presence of serum in the medium remained healthy for at least a week in culture.

In contrast, when asbestos fibers were added to macrophages in buffered saline medium lacking serum, marked cytotoxic effects rapidly occurred. Thus, when mouse macrophages were exposed to 100µg/ml medium of IUCC standard chrysotile, 90% of cells had lost the capacity to split fluorescein diacetate within two hours. Release from cells of protein, including the cytoplasmic enzyme lactate dehydrogenase, was marked within one hour of exposure to asbestos, whereas release of acid phosphatase and β-glucuronidase was delayed until later. Cinephotomicrography showed sudden cessation of movement and sometimes blebbing of the cell membrane. Hence, it seems highly probable that this type of rapid cytotoxic reaction is due to a direct interaction of asbestos fibers with cell membranes, analogous to hemolysis. The analogy is strengthened by the observation in my experiments that chrysotile is the most effective of all the IUCC samples in

early cytotoxicity, followed by crocidolite, amosite, and anthophyllite
in that order. This is in contrast to
the rather surprising report of Parazzi et al[50] that the rapid early cytotoxic reaction of crocidolite is greater than that of chrysotile.

In any case, it is doubtful whether
the rapid cytotoxic reation produced
by asbestos fibers in media lacking
serum or bronchial washings has
much physiological relevance. I have,
therefore, examined the long-term
effects on macrophages and mesothelial cells of exposure to fibers that
are mostly of ingestible size in medium containing 10% fetal calf serum.
Such fibers are taken up into vacuoles that become secondary lysosomes, as shown by acid phosphatase
and other marker enzymes. It is already clear that all types of asbestos
dusts are much less toxic to macrophages and other cells than are silica
dusts. However, it would be wrong to
conclude that asbestos dusts are entirely devoid of cytotoxic potential.
Macrophages exposed to chrysotile
(20 to 50μg/ml IUCC standard)
often show after two or three days
in culture very large lightly staining
vacuoles in their cytoplasm (Fig 3).
Some cells show pyknotic cytoplasm
and nuclei and about 20% are no
longer viable. Hence, there was certainly some cytotoxic reaction, as
well as other abnormalities. One of
the most striking was the development in mesothelial cells, and to a
lesser extent in macrophages, of
brown pigmented autofluorescent
granules in the cytoplasm; this was
greatly accentuated after ingestion
of asbestos, especially crocidolite.
Perhaps the iron favors lipid per-
oxidation.

Apart from formation of iron-containing pigment, already mentioned,
the most striking ultrastructural
change in asbestos-treated cells is
the development of an electron-dense
matrix surrounding the asbestos particles in secondary lysosomes. This is
not seen in silica-treated cells and
could provide a protective layer between the asbestos particles and the
vulnerable membrane surrounding
the secondary lysosome. This may, in
fact, explain why asbestos is less toxic than would be expected from its
capacity to react with membranes, as
shown by hemolysis.

Although much remains to be
learned about the effects of asbestos
on cells, certain important differences from effects of silica are already apparent. One is the failure
of ingestion of many asbestos fibers
because they are too long. Another is
the obviously weaker cytotoxicity of
asbestos than silica, although some
forms of asbestos, especially chrysotile, are certainly more toxic than inert dusts such as anatase or diamond
dust. The slowly progressive damage
to macrophages so produced may be
an important factor in the fibrotic
reaction to asbestos. Also observed in
asbestos-treated cells is lipid per-
oxidation and accumulation in vacuoles or iron-containing pigment; this
contributes to the formation of asbestos bodies in vivo.

References

1. Stishov SM, Belov NV: Crystal structure of a new dense modification of silica. *Dokl Akad Nauk SSSR* **143**:951, 1962.

2. Bohn E, Stover W: Isolation of coesite and stishovite as natural minerals. *Neues Jb Miner Mb* **5**:89, 1966.

3. Preisinger A: Structure of stishovite, high density SiO_2. *Naturwissenschaften* **49**:345, 1962.

4. Deer WA, Howie RA, Zussman J: *Rock-Forming Minerals*. London, Longmans Green & Co Ltd, 1962.

5. Marks J: The neutralization of silica toxicity in vitro. *Brit J Industr Med* **14**:81, 1957.

6. Marks J, Nagelschmidt G: Study of the toxicity of dust with use of the in vitro dehydrogenase technique. *Arch Industr Health* **20**:383-389, 1959.

7. Vigliani EC, Pernis B, Monaco L: Study of the action of quartz particles on macrophages in vitro, in Davies CN (ed): *Inhaled Particles and Vapours*. London, Pergamon Press Inc, 1961, p 348.

8. King EJ: Volubility theory of silicosis. *Occup Med* **4**:26, 1967.

9. Curran RC, Rowsell EV: The application of the diffusion-chamber technique to the study of silicosis. *J Path Bact* **76**:561, 1958.

10. Vigliani EC, Pernis B: Immunological aspects of silicosis. *Adv Tuberc Res* **12**:230, 1963.

11. Voisin GA, Collet A, Martin JC, et al: Propriétes immunologigues de la silice et de composés du beryllium: Les formes solubles comparées aux formes insolubles. *Rev Franc Etud Clin Biol* **9**:819, 1964.

12. Allison WC, Harington JC, Birbeck M: An examination of the cytotoxic effects of silica on macrophages. *J Exp Med* **124**:141, 1966.

13. Nadler S, Goldfischer S: The intracellular release of lysosomal contents in macrophages that have ingested silica. *J Histochem Cytochem* **18**:368, 1970.

14. Cohn Z, Hirsch JG, Wiener E: Lysosomes-endocytosis: Cytoplasmic granules of phagocytic cells and the degradation of bacteria in De Reuck AVS, Cameron MP (eds): *Lysosomes: Ciba Foundation Symposium*. Boston, Little Brown & Co, 1963, p 126.

15. Beck EG, Bruch J, Sack J: Beobachtungen über die Morphologie der Staubphagozytose in vitro. *Ergebn Unters Geb Staubund Silikosekek Steinkohlenbergbau* **6**:131, 1967.

16. Beck EG: Die Reaktion in vitro gezuchter Zellen auf partikelformige Luftverunreinigungen und hochpolymere Stoffe. *Forschungsberichte Landes Nordrhein-Westfalen* No 2083, pp 125.

17. Bruch J: Ein elektronenmikroskopischer Beitrag zum Frühstadium der Silikose. *Fortschr Staublungensforsch* **2**:249, 1967.

18. Bruch J, Otto H: Electronenmikroskopische Beobachtungen an Alveolarmakrophagen in der Rattenlunge nach Quartzstaubinhalation. *Ergebn Unters Geb Staub-Silikosekek Steinkohlenbergbau* **6**:141, 1967.

19. Collet A, Martin JC, Norman-Reuet C, et al: Recherches infra-structurales sur l'evolution des macrophages alveolaries et leurs reactions aux poussieres minerales, in Davies CN (ed): *Inhaled Particles and Vapours*. Davies

Pergamon Press Inc, 1967, vol 2, p 155.

20. Ruttner JR, Grieshaber E, Vogel A, et al: Phagozytoze und Fibrillogenese im peritonealen Staubknotchen electronen mikroskopische befunde bei Kaolin-induzierten knotchen. *Fortschr Staublungensforsch* **2**:89, 1967.

21. Comolli R: Cytotoxicity of silica and liberation of lysosomal enzymes. *J Path Bact* **93**:241, 1967.

22. Sakabe H, Koshi K: Preventative effect of polybetaine on the cell toxicity of quartz particles. *Industr Health* **5**:181, 1967.

23. Stalder H, Stober W: Haemolytic activity of suspensions of different silica modifications and inert dusts. *Nature* **207**:874, 1965.

24. Nash T, Allison AC, Harington JS: Physicochemical properties of silica in relation to its toxicity. *Nature* **210**:259, 1966.

25. Rotman B, Papermaster BW: Membrane properties of living mammalian cells as studied by enzymatic hydrolysis of fluorogenic esters. *Proc Nat Acad Sci USA* **55**:134, 1966.

26. Schlipköter HW, Dolgner R, Brockhaus A: Ein Beitrag zur Therapie der experimentellen Silikose. *German Med Monthly* **8**:509, 1963.

27. Schlipköter HW: Die Wirkung von Polyvinylpyridin-N-oxid (P204) im Tierexperiment. *Arbeitsmed Sozialmed Arbeitshyg* **4**:133, 1967.

28. Beck EG, Bruch J, Brockhaus A: Alteration of the cytotoxic effect of quartz on mouse fibroblasts (strain L) by polyvinylpyridine-N-oxide. *Z Zellforsch* **49**:568, 1963.

29. Schlipköter HW, Beck EG: Observations on the relation between quartz cytotoxicity and fibrinogenicity while testing the biological activity of synthetic polymers. *Med Lavoro* **56**:485, 1965.

30. de Duve C, Wattiaux R: Functions of lysosomes. *Ann Rev Physiol* **28**:435, 1966.

31. Grundmann E: Experimentelle Untersuchungen über die zellulare Speicherung des Polyvinyl-pyridin-N-oxid. *Fortschr Staublungensforsch* **2**:223, 1967.

32. Beck EG, Boje H: Zytologische Untersuchungen über die Wirkung vol Poly-(2)vinylpyridin-N-oxid in der Zellkultur. *Fortschr Staublungensforsch* **2**:231, 1967.

33. Bairati A, Castano P: Studio al microscopio elettronico della localizzazione, nel fegato dei topi, di polimeri sintetici citoprotettivi nei confronti della silice (poli-vinipriridino-N-ossido e poli-p-dimetilamminostirolo-N-ossido). *Med Lavoro* **59**:81, 1968.

34. Dehnen W, Fetzer J: Über die Wirkung von Polyvinylpyridin-N-Oxid (P204) au die Stabilitat isolieter Ratten leberlysosomen. *Ergebn Unters Geb Staub-und Silikosekek in Steinkohlenbergbau* **6**:161, 1967.

35. Holt PF, Lindsay H, Beck EG: Some derivatives of polyvinylpyridine 1-oxides and their effect on the cytotoxicity of quartz in macrophage cultures. *Brit J Pharmacol* **38**:192, 1970.

36. Munder PG, Modolell M, Ferber E, et al: Phospholipids in quartz-damaged macrophages. *Biochem Z* **344**:310, 1966.

37. Heppleston AG, Styles JA: Activity of a macrophage factor in collagen formation by silica. *Nature* **214**:521, 1967.

38. Vorwald AJ, Delahant AB: The influence of silica on the natural and acquired resistance of the tubercle bacillus. *Amer Rev Tuberc* **38**:347, 1938.

39. Hart PDA, Aslett EA: Chronic pulmonary disease in South Wales coalminers: I. Medical studies. *Med Res Counc Spec Rep Series* **243**:186, 1942.

40. *Miners' Chest Diseases Treatment Centre Report.* London, Medical Research Council, 1967.

41. Cesa-Bianchi D: Staubinhalation und Lungen Tuberculose. *Z Hygiene Infektionskrank* **73**:166, 1963.

42. Vorwald AJ, Dworski M, Pratt PC, et al: BCG vaccination in silicosis. *Amer Rev Tuberc* **62**:455, 1950.

43. Policard A, Gernez-Rieux C, Taquet A, et al: Influence of pulmonary dust load on the development of experimental infection by *Mycobacterium kansasii. Nature* **216**:177, 1967.

44. Pernis B, Paronetto P: Adjuvant effect of silica (tridymite) on antibody production. *Proc Soc Exp Biol Med* **110**:390, 1962.

45. Wilkinson PC, White RG: The role of mycobacteria and silica in the immunological response of the guinea-pig. *Immunology* **11**:229,1966.

46. Mackaness G: Cellular immunity, in van Furth R (ed): *Mononuclear Phagocytes.* Oxford, Blackwell Scientific Publications, p 461, 1970.

47. Allison AC, Hart PDA: Potentiation by silica of the growth of *Mycobacterium tuberculosis* in macrophage culture. *Brit J Exp Path* **49**:465, 1968.

48. Macnab G, Harington JS: Haemolytic activity of asbestos and other mineral dusts. *Nature* **214**:522, 1967.

49. Secchi GC, Rezzonico A: Hemolytic activity of asbestos dust. *Med Lavoro* **59**:1, 1968.

50. Parazzi E, Pernis B, Secchi GC, et al: Studies on "in vitro" cytotoxicity of asbestos dust. *Med Laboro* **59**:561, 1968.

51. Davis JMG: The effects of chrysotile asbestos dust on lung macrophages maintained in organ culture: An electron-microscope study. *Brit J Exp Path* **48**:379, 1967.

Political Aspects of Air Pollution Control

The Federal Response
to Air Pollution

RAYMOND D. COTTON, JD, SM Hyg

Although this book [Vanishing Air] *brings to light many hitherto undisclosed facts and events, its most significant contribution is its analysis of the collapse of the federal air pollution effort starting with Senator Edmund Muskie and continuing to the pathetic abatement efforts and auto pollution policies of NAPCA* [National Air Pollution Control Administration].[1]

Ralph Nader

IN the above statement, Mr. Nader in quite harsh terms was referring to the Air Quality Act of 1967[2] (which amended the clean Air Act[3]) and to the performance of the National Air Pollution Control Administration (NAPCA) under these laws. Fairness demands that the underlying philosophy of the Clean Air Act, as amended, be recalled. Section 101 of that statute declares the following Congressional findings:

(3) that the prevention and control of air pollution at its source is the primary responsibility of States and local governments; and

(4) that Federal financial assistance and leadership is essential for the development of cooperative Federal, State, regional, and local programs to prevent and control air pollution.[4]

The stated Congressional intention of leaving primary responsibility in the hands of the states and local governments is unmistakenly set forth in the above "findings." Even when the federal government is given a role, it must be "cooperative." In a sense, therefore, Mr. Nader may have been criticizing NAPCA for being too faithful to the program called for by Congress.

Recently, there have been significant legislative changes which bring more power to both the federal government and to the consumers. Two significant and new provisions which are the focus of national attention will be discussed.

Motor Vehicle Emissions Standards

One of the weaknesses of the Clean Air Act of 1967 was that, while it made general provision for the establishment of motor vehicle emission standards, it specified no date by which the manufacturers had to comply.[5] Furthermore, it did not permit NAPCA to select engines for testing; such selection was to be made *by the manufacturer*. NAPCA had no legal authority to enter the plant or the production line for purposes of inspection.

The new law is a good deal more detailed on these and other points. In effect, Section 202(b) requires that by 1975 carbon monoxide and hydrocarbon emissions from automobiles must be at least 90% less than those allowed by 1970 standards. It is further required that by 1976 nitrogen oxide emissions must be at least 90% below the 1971 standards. The Administrator of the Environmental Protection Agency (EPA) may permit a one-year extension of each deadline, but politically it would probably be almost impossible for him to do so.

The enforcement provisions now permit the agents of EPA to enter manufacturing plants for, as the statute says,

. . . the purpose of conducting tests of vehicles or engines in the hands of the manufacturers, or to inspect . . . records, files, papers, processes, controls, and facilities used by such manufacturer in·conducting tests under regulations of the Administrator [of EPA].[6]

For each violation of the standards set by EPA, the maximum fine is $10,000. Clearly, this could mean $10,000 per automobile in an egregious case.

Citizen Suits

A new section was added to the Clean Air Act of 1967 by the Clean Air Amendments of 1970, which would allow civil suits by individuals against violators of established emission standards.[7] Even government agencies may be sued for such violations. In fact, there is a specific provision which would permit an individual to sue the Administrator of the Environmental Protection Agency in order to force him to perform his duty under the law. Could it be possible that a Democratic Congress did not trust a Republican Administrator?

All such suits are required to be brought to the Federal District Court. This is the lowest court in the Federal system, but it is of high legal quality. Trial by jury is an option of the defendant, and the losing party could have two appeals, one to the Federal Circuit Court of Appeals and then to the US Supreme Court.

It is interesting to note that this section attempts to encourage legitimate citizen suits against polluters and to discourage frivolous legal actions by providing the Court with power to award the costs of attorney and witness fees to any party of the litigation. Such authority is rarely permitted in courts of this country and usually must be granted by specific statutory language.

Conclusion

With the massive amendments and additions to the Clean Air Act by the Clean Air Amendments of 1970, with the transfer of NAPCA to EPA, with the appointment of an extremely capable Administrator for EPA, and with strong apparent support in Congress for EPA, one may easily predict great progress in air pollution control. Two of the more controversial parts of the Clean Air Amendments of 1970 have been discussed here.

References

1. Nader R: Introduction, in Esposito JC: *Vanishing Air.* New York, Grossman Publishers, pp vii-viii.

2. *Air Quality Act of 1967,* 42 USC, 1857.

3. *Clean Air Act of 1963,* 42 USC, 1857.

4. *Air Quality Act of 1967,* 42 USC, 1857a.

5. *Air Quality Act of 1967,* 42 USC, 1857f-1.

6. *Clean Air Amendments of 1970,* Public Law 91-604. US Congress, House of Representatives, 91st Congress, second session, Dec 31, 1970, sec 206(c), pp 20-21.

7. *Clean Air Amendments of 1970,* Public Law 91-604. US Congress, House of Representatives, 91st Congress, second session, Dec 31, 1970, sec 304(a), pp 31-32.

NEW ATTITUDES TO AIR POLLUTION—
THE TECHNICAL BASIS OF CONTROL

R. S. SCORER

Abstract—Technical knowledge is needed to establish priorities for action. It is important to survey each situation to discover which of the practicable means available will be the most effective, and a procedure for conducting the survey is indicated in an example—the basis of the 1956 U.K. Clean Air Act is examined.

The technical knowledge needed lies as much in the fields of Local Authority administration, economics, and public relations, as in technology and pure science, and a more effective dialogue is needed between politicians and those working in scientific fields. This dialogue requires personal contact and a more widespread proficiency in the art of briefing and convincing the critical layman on both sides.

After considering legislative and other methods, control by practitioners (like the Alkali Inspectorate) is recommended in preference to enforcement through the courts. Admittedly it is consideration mainly of British air pollution problems that leads to this conclusion and it is recognized that technical details of a legal, geographical, economic, social, or historical origin may make different methods appropriate in other countries. Costing, and control by fiscal methods are not likely to be effective.

At the international level water pollution is far more serious than air pollution, which must not be relieved by creating more water pollution. Some of the outstanding air pollution problems are considered. The most important is car exhaust, and clean exhaust zones are recommended in preference to country-wide restrictions. Globally the two greatest problems are the increase in CO_2 and the production of particulate haze: these work to change the climate in opposite directions.

The pressure for economic growth in the most advanced countries conflicts with the preservation of the environment. Already the richest nations make more than a fair share of pollution, for the result would be catastrophic if every person made as much, on the average, as the average in the richest countries. This fact alone means that the economic forces which direct the development of our society must be changed, and to make this possible many of our conventions and social criteria need to be modified.

1. INTRODUCTION

UNTIL the present century it had been possible for man to dispose of unwanted by-products of his activities in the immediate environment where they were digested, almost incidentally, by natural biological activity. Today the areas where the by-products are not being satisfactorily digested are spreading: the ecological balances are overturned and destruction is progressive and cumulative.

The prospect to be faced, therefore, is that this destruction will soon extend to such a large fraction of the earth's surface as to make a drastic, and perhaps catastrophic change in our way of life necessary. The challenge is to take note of the trends and warnings and see how our economic and social systems and attitudes must be changed if we are to change the direction of civilized evolution before it brings disaster upon itself.

Each of us harbours personal views about the responsibilities of individual and collective personal behaviour. The laws, customs, and criteria by which Western society lives are the results of experimental evolution, with plenty of time for mistakes and minor catastrophes and a gradual change in habits of conduct. An increase in social tolerance and personal wealth has been coupled with less fundamental dogmatism. But it is mainly our individual economic independence which protects us from people with greater power. The Communist and newly developing nations are attempting to establish societies motivated by different criteria, but in none of the three worlds are the changes to new social values primarily determined by the need to limit growth of population and industrial activity in the near future.

We have to appreciate that controls and limitations of a new kind will have to be imposed either by dictation from above, or by the evolution of a new consensus. Emotional prejudices which have played a valuable role in social structure till now will be projected by those with dogmatic beliefs into the role of solutions for our new problems. For this reason it is desirable to examine the technical basis of the problems and some of our present methods of dealing with them. This essay gives only preliminary assessments that some existing methods are good and others are bad. But any apparent dogmatic finality should be seen as designed to upset prejudices which are inhibiting progress of thought, and to make the next steps towards more beneficial methods easier; it is not intended to express a fully considered ultimate faith. For this reason it is hoped that those who disagree will argue by offering more attractive proposals rather than by attempting to fault the logic in arguments which really stem from differences in knowledge which produce differences of opinion and taste. New knowledge alters preferences. If we are to succeed in avoiding a doctrinaire dictatorial bureaucracy we must recognize that action will in fact be a muddle of compromises and experiments.

This essay is a revised version of one written for the Conference on Pollution organized by the British Liberal Party in November 1970, and to be published in the volume reporting that conference. The author is a member of the Labour Party.

2. DIALOGUES IN A TECHNOLOGICAL SOCIETY

The division of labour is more striking as technology becomes more complex. Designers, managers, accountants, salesmen, planners, machine or assembly line operators, maintenance men, construction engineers all have rather narrowly defined roles in modern industry. Scientists are regarded as professional dispensers of skills and knowledge which may be exploited by organizations, but only in rare cases are scientists deliberately involved in formulating the policies which determine how their science and technology shall be exploited. They have usually been fairly content to be given the chance to develop their techniques, and their training has tended to make them amoral.

To some extent the amorality of science has made non-scientists think of scientists as unpolitical or perhaps as slightly irresponsible and unpractically idealistic in politics. Outstanding examples of individuals typify this and sustain this picture of the scientist. In fact most scientists are very concerned about the direction in which technological man is moving but they tend to have most of their time and energies consumed in the pursuit of technical aspects of their science.

Very few enter the business of politics: not so lawyers and economists who abound in the political world.

But many modern problems are not predominantly to do with the economy or with law. In education, medicine, the provision of public services such as electric power and drainage, exploitation of mineral resources, and large constructional developments scientific experts are becoming more involved in decisions which are important politically. Of course there are legal and financial consequences but these are secondary.

In air pollution we cannot separate the technical aspects (such as medical, chemical, and meteorological) from the rest, and so it is inevitable that scientists cannot be kept out of the councils where political decisions are made. If scientists were more deliberately involved and required to participate in solving legislative and administrative problems it would affect the way they conduct their scientific investigations. More particularly it would enable them to proffer more efficiently the relevant information and to volunteer views on the implication of proposals.

The division of work is only safe if there is a genuine dialogue between the experts. It is not always appreciated by non-scientists that scientists do not really spend their time asking questions and answering them. Indeed this is not by any means properly understood by scientists themselves, for in writing accounts of their researches they often pretend that they asked questions and answered them—the same questions. Many have lately got into the habit of introducing their work by such statements as "we asked ourselves the following questions: . . . " or "The problem posed was as follows: . . . ". In fact the question is usually formulated after the answer has been found. Seldom is a question answered in its original form, for this often turns out to be silly or even nonsensical, and a continual evolution of objectives takes place while the work goes on and more knowledge is gained. The dialogue between the scientists and others must recognize this.

There is no special quality of science in this respect, for social, political and economic objectives also evolve as problems are worked upon and solutions experimented with. Much of the evolution is simply the discarding of old prejudices, or concepts, and this is what causes the old questions to become rather silly.

We shall see, in this essay, how technical knowledge affects choice of legislation and legislative methods, and how technical factors of a local nature influence the suitability of certain kinds of legislation. Much of the evolution which makes different solutions to problems appropriate in different countries is of a social or political nature, but much of it is equally geographical or meteorological, and in discussing different means of air pollution control all these have to be understood: none can be ignored.

There is a task of a very modern kind here: the scientist must learn to talk about his contribution to solving problems in a way that enables non-scientists to appreciate the issues which he has been facing. Economists attempt a similar task to some extent, as do lawyers who wish the public to feel a sympathy for their efforts (justice must seem to be done). But chemistry and mathematics tend to become more technical and remote: there is such a wealth of detailed structure in science upon which the outcome of any piece of work depends that it seems almost impossible for the non-scientist to penetrate it and it looks as if he must simply be told what the "answer" is.

If that view be accepted it is equivalent to recognizing that scientists operate at an intellectual level beyond the reach of ordinary people. But scientists often have difficulty in communicating even with each other because they do not have each other's specialist knowledge and we do not rate a chemist as cleverer than a mathematician or vice versa because he cannot be understood by him.

It is a matter of opinion whether this situation should be accepted. It is certainly most unhealthy from the political point of view to have blockages in communication between people who are all contributing to a final decision. Consequently it is important that scientists should assume the task of making themselves intelligible to ordinary people. It is an art they must learn, and they will learn it only by practice. They will learn it more effectively if invited to practise it and to become part of the decision-making team, and learn to contribute to discussion of the social aspects of the problem. This is not a plea for involvement for its own sake, nor for the right of participation, however valid those arguments; it is simply a view of how the efforts of the scientist can best be directed in the service of humanity.

There is nothing new in the idea, but it has not always been practised. It is much more desirable than the employment of scientists for political objectives or in war without the scientist having the chance to support or oppose the cause, and it would equally be deplorable if scientists automatically entered into the service of the highest bidder.

The need for involvement and the development of intelligible communication between scientists and others is well illustrated by the difficulties facing legislators when scientific advice is contradictory, as it was in the case of cyclamates. The futility of courtroom procedures for finding the truth out of a welter of conflicting evidence is often exemplified at public enquiries where "experts" are obtained by both sides to support their case. These experts (and I speak from personal experience) are often cross-examined by advocates whose sole purpose is to discredit them. The questions asked and suggestions made are often quite silly from the scientific point of view. The expert may be able to make the advocate look small by comparing his understanding with that of an ignorant student, but it is more important that he should be able to explain his case genuinely, and this means by making his understanding clear first to his own counsel and then to Mr. Everyman.

This is only achieved by a two-way effort. I do not accept the view of one politician* "I think that politicians have the right to expect more precise scientific judgments" by which he meant that the scientists should give clear and unequivocal answers to the politicians' questions, and on the basis of these answers the politicians could act. The danger lies in the wrong questions being asked and the answers, which must necessarily be a grossly simplified version of the real situation, being misunderstood or applied in a modified context which was not specified in the original question.

The exclusive political viewpoint is typified in the sentence (*loc. cit.* p. 111) "I don't like the thought that cars are being made in this country for export to America which are being forced to install exhaust systems which will reduce the amount of pollution they put into the atmosphere but the cars that are being sold to the British public will not have this system". As we shall see, the technicalities which apply in America may not apply here and the law in America requiring the system may be based on the answer to the wrong question or it may be a law of a kind which tends to be introduced in the American political system. The "thought" which is not liked is derived from inappropriate prejudices. The scientist must understand the politics and the politician the science.

---

* University of Edinburgh—*Teach-in on Pollution* 6 March 1970, p. 128, Science Studies Unit Edinburgh University.

3. CONTROL OF AIR POLLUTION BY FISCAL METHODS

If we could assign the cost of the damage done by air pollution to the various sources it might be possible simply to charge the cost to the source. The economics of the situation would then cause the polluter to seek means to avoid making the nuisance, but only if their costs were less than the cost of the damage would they be used. The philosophy is that it is better to suffer a little damage and get our accounting right than to make the pollution illegal and employ excessively expensive means to prevent it. The criterion for choice of one alternative is the total advantage to the community.

We shall see later how far it is possible to go in assigning the damage to particular sources and what are the limitations to costing all the damage. There is, however, a more important objection to the use of such methods, namely that they enable polluters to buy the right to pollute. Society would not know in advance whether a polluter would choose to pay the penalty or avoid making the pollution. The law should be framed so as to achieve its objective for certain.

If it were proposed to raise the levy on polluters until they were driven out of business they might as well be prohibited from the beginning. If we pursue the original objective of promoting the choice which is to the greatest total advantage to the community, then society's choice must be forced upon the polluter, otherwise he could simply raise his prices and his customers would have to pay for his decision. The supposition that he would be forced out of business by competitors is not generally correct in a modern affluent society, which is the kind of society in which pollution is the most serious problem. There are many monopolies particularly in the U.S.A., such as the "Public Service" electric supply companies which cannot be controlled by crude economic pressures, and in modern economics a very considerable part of the total expenditure is controlled by governments. In the U.S., government expenditure, over which the forces of the market do not operate, is of the order of a quarter of the total, and covers not only ordinary civil and military expenditure but also vast projects sponsored by public civil or military agencies, such as N.A.S.A. Rising costs do not determine policy in these cases although they affect the rate at which the projects are carried out.

But even supposing that fiscal methods could be applied within a large civil sector of the economy, covering most commercial undertakings, the logic of economics would not necessarily be overriding. People make decisions according to their ability to comprehend the issues: it is already difficult enough to make decisions not directly connected with production such as staff amenities, conventional methods of reducing pollution such as chimneys, and expenditure on advertising and public relations. It is probable that the levy on polluters would be regarded as another tax to be paid reluctantly and forgotten about as quickly as possible. To avoid it would be a more complicated procedure and just not worth the worry. It might be argued that in such a case it was, *in toto*, less trouble for the community to allow the pollution to continue than to abate it, but it would only be trouble this year or this month, not total trouble in the long-run that would be considered.

The emotional aspects are illustrated in the extreme by the smoking problem. Increased taxes on cigarettes have almost no effect on consumption, whereas a prohibitive tax on cigarette advertising might be quite effective in reducing the number of people who take up smoking. For social and psychological reasons to do with the sedative and addictive aspects of smoking, it is useless to pass the cost of the damage on to the consumer; it is already paid mostly by him since he suffers it almost alone, although the subsequent medical costs and loss of his work to the community are large. The high cost of illegal drugs on the black market does not deter the addict but merely makes him more ruthless in his methods of obtaining the money. In normal commerce there do exist psychological pressures which override crude accounting, and some are good in their effects; and it cannot be assumed that financial pressures will override other influences. The failure of the "accounting" philosophy is illustrated well by the spoilation of land. Coal mine owners were able to place tips of waste material on low priced agricultural land. The tips have remained, and two or three generations have suffered the ugliness of the result. The cost of restitution will now be enormous. It is important therefore that the accounting system should not seem to reduce costs merely by postponing payment, while actually increasing them.

The complexities of using accounting methods become more obvious in the case of rats, bottles, and cartons. When the Lindsey County Council offered "a shilling a tail" for rats one farmer who had collected tails over the years, dusted them over and cashed them for several pounds. Others would let a female go free, but tailless, in the "hope" that it would breed more, while some actually took up breeding them in cages. We are now beginning to appreciate the public nuisance of "no-return" bottles, and the enormous increase in recent years of packaging of all kinds which accumulates in homes. The proposal that salvage should be subsidized always faces a dilemma: if it is made too profitable the waste material (bottles or packaging etc.) is likely to be created in response to the payments offered. The only way out is to make the suppliers of bottled products obliged by law to remove returned bottles. Unless something of this kind is done soon every civilized community will find itself extracting more minerals from the ground and turning more wood into paper and oil into plastics and at the same time creating great waste dumps of the used products. The re-use of scrap and waste materials is an essential in a populous country with a "high standard of living", i.e., a large material turnover. Nevertheless we shall see later that a reduction in consumption is also imperative.

In the context of air pollution, this principle must mean the avoidance of the emission of it. In the context of water pollution, we already have a public service for the treatment of sewage and water is used several times during its passage down a river. Unless other means are soon found the salvage of glass, paper, plastics, etc., will have to be undertaken as a public service. The question arises how much we should allow an industry to make use of such a public service to avoid responsibility for the consequences of its method of marketing its products. One cannot help sympathizing with the view that fiscal means should be used to make the responsibility appear in the company's accounts, but that would probably be inefficient in the case of material salvaged, although it could be used to enforce as much re-use of bottles, for example, as possible. This is a very real problem, for if we allow the artificial rules of an expanding economy to dominate the age of affluence milk bottles will disappear and our households will accumulate piles of plastic coated paper cartons which cannot be salvaged, our newspapers will be three times as thick, inflated by unread advertisements and tedious magazine material and the forests of the world will be pillaged to destruction and their ability to absorb CO_2 greatly reduced.

These are not trivial issues: they are all part of the choking of the affluent society in its own refuse because it occupies too much of the world. The doom predicted by

Malthus for an exponentially growing population will apply soon to a linearly growing economy. The expanding economy must not be measured simply in material turnover. If it is to be measured in terms of money at all the accounting must include some sort of cost-benefit analysis of pollution. We now examine the possibility of this.

4. IS A VALID COSTING POSSIBLE?

The issue here is not so much whether a cost analysis can be made but whether it is valid. In order to set a value on the tulips in the park we find out how much is spent on them. This is then alleged to be "the value that society sets on them". This procedure is false because in general society does not know the cost at the time decisions to spend are made. It is similar to the phenomenon of people believing that tax is a disincentive to work when even in their own individual cases it isn't because they do not know how much tax they pay. If they were asked to say whether a "penny rate" was appropriate for the tulips they would reply according to what they thought a penny rate meant (a penny per head, per rate payer, per annum, or once for all, and so on). The fact is that people do not set financial values on these things but appoint other people (the employees of the local authorities or in some cases the councillors themselves) to make the decisions for them. The low polls at local elections show that even that aspect concerns only a very few people. The people appointed do not actually value the public services one against the other. There is always a precedent for the proportion of the rates allocated to different purposes and departures from these precedents result either from the pressure of necessity (to repair a bridge, to construct a by-pass or a new sewage plant) or from the personality of an individual advocating a higher expenditure for a particular purpose which he believes to be very worthwhile.

The second difficulty in setting a monetary value on an amenity is that the value a community gives it depends on its experience or knowledge of it. When we spend a quarter of a million pounds on a painting "for the nation" it is, of course, not for the nation, but for a very small minority who have influence in the quarter where the decision is made, and their propaganda has caused the rest of the community to accept the decision. In many towns, where the air is very dirty, nothing is done to institute smoke control either because the existence of the smoke has not been recognized (as in the case of Epsom, which is reluctant to accept its description as a "black area" in spite of the smoke measurements made there) or because the advantages of clean air are not appreciated since they have never been experienced, and the polluted air has come to be regarded as the normal order of things. We accept much of our environment uncritically through familiarity or because we have not actually judged it by external criteria. Sir Roger Duncalf, the deputy chairman of the Beaver Committee appointed after the great smog of 1952, used frequently to say that until he joined the committee he had not noticed most of the pollution around him, even in his own house. This is normal. Pollution is accepted like the weather.

When clean air has actually been experienced, and the differences between controlled and uncontrolled air pollution observed as such, the value set upon it by the community increases greatly. One can measure the value unconsciously given to clean air by house buyers who react to the squalor of a district by paying lower prices, but they may equally be reacting to the value set by people in a neighbourhood on new paint. The sight of one newly painted house often induces other householders to paint

128

theirs too, although they had not previously thought of it, and the decision to paint is not made by consciously forfeiting an alternative extravagance.

One method of circumventing this problem is to make use of more than one currency, exchange between them not normally being possible. In so far as we can subsequently attribute costs to various activities such as tulips in the park and subsidies for smoke free heating appliances, we are determining what rate of exchange resulted from policy decisions, although the policy might have been determined without reference to the relative costs of the activities, all of them having been placed within the budget beforehand.

It can be argued that we ought to do the exercise of considering the relative costs in advance: it is only prudent to do so in order to make best use of our resources. If we carry this as far as possible it tends to devalue our aesthetic judgments. To agree to a complete cost analysis in advance requires that our purposes shall either be formulated without reference to finance, in which case the analysis simply serves to help us find the most effective method of achieving our purpose, or it requires a subordination of the policies to financial considerations. If the financial considerations were untainted by contemporary tax systems and conventions and laws of property and the way banking pressures operate in our particular society, the cost could be accepted as a nearly absolute measurement of something. But they are not. The legal set up we have is the one that has produced pollution problems and the mere fact that fiscal methods for the control of pollution have been considered implies that legal means can be used to alter the appearance of our account books.

The only attempt I know of to simulate the real situation in a simple model is the game "Careers". Like "Monopoly" it is a mixture of luck and choice. There are three currencies—wealth, fame, and happiness—and the player decides the proportions of these in his ambition before the game begins. When, from time to time, he has the opportunity to enter upon a career he decides whether to do so according to what it offers in each of the three currencies. During the course of the game limited opportunities arise to "buy happiness" and so on, but the circumstances for the interchange of the currencies are very restricted. The game does not provide any solutions to our problem, but merely states the situation in an interesting manner.

The prospects are, therefore, that any cost analysis is likely to be restricted to comparisons between entities which are very similar or to comparisons between very dissimilar entities only in very special situations. Most of the time choices have to be made on aesthetic or political grounds and a cost comparison is either not possible, excluded by circumstances (such as pre-allocated budgets), or only possible afterwards as a "costing" of decisions within the contemporary legal framework.

Costing methods are sometimes bound to fail. For example the problems of paper production and disposal are large in the pollution field. Sweden suffers most of the pulp production pollution problems caused by the use of paper in Britain while Britain has a major refuse problem. At present they can only receive quite separate treatment, but ultimately the only solution is a reduction in paper consumption, and this can only be achieved painlessly by international arrangement.

In problems which are international the prospect of solution is greatly increased if an industry can form an organization which can speak to governments on behalf of the whole of it. The oil industries of Western Europe have formed such an organization (CONCAWE) which must surely be followed in other polluting industries such as paper and cement (and perhaps even some agricultural industries, such as sugar and wheat, which are major sources of smoke in the world). The creation of such organizations would hasten the day when world production can be stabilized at a socially desirable total annual tonnage.

5. COST OF AIR POLLUTION AND THE DAMAGE ATTRIBUTABLE TO DIFFERENT SOURCES: THE EFFECT OF THE CLEAN AIR ACT (1956)

Some progress can be made in cost analysis of the harm done by different polluters. The objective is to estimate the probable effectiveness of various remedies in order to ensure that our efforts will be worth while. In this essay I shall not be particularly concerned to justify the scientific judgments which have to be made: rather my purpose is to emphasize where they are necessary, and where the calculation lacks precision. It will appear also that any extension of the method to make a comparison between air pollution and water pollution, smoking, and noise is almost impossible, and that to determine our priorities in protection of human health and the environment we have to consider time and other factors.

First we divide the sources of pollution into categories. These must be of comparable importance in the community, they must be groupings for which it is possible to legislate in a discriminatory way, and each group must be more or less homogeneous, otherwise it will not be possible to assign definite different characteristics to the groups. The first consideration makes it clear that the groups chosen for the whole country could be different from those chosen for a small area. For example steel works are an obviously appropriate category for the Sheffield–Rotherham conurbation but not for Manchester, while they could be included in "metallurgical and other similar industries" (perhaps including glass and chemicals) for the country as a whole. TABLE 1 gives the categories chosen for a preliminary analysis of the whole of Great Britain.

TABLE 1. CLASSIFICATION OF SOURCES OF POLLUTION

1. Electricity generating stations
2. Other major industrial plants and complexes such as oil refineries, car factories, chemical works, steel works, cement works, etc., which use large amounts of fuel and mostly have tall chimneys
3. Smaller industrial plants such as brick works, foundries, mills, collieries, etc.
4. Offices, hospitals, apartment blocks, hotels, and other large buildings with heating plants
5. Private homes
6. Road traffic
7. Trains, ships in port, aircraft at airports

The crudity of the analysis makes it undesirable to have too many categories. Criticisms are easily made of any classification and only serve a purpose if they improve the quality of the conclusions drawn. Cases can easily be found which do not easily fit into the classification, but that does not affect the general nature of the guidance which emerges.

Secondly we have to determine the main independent features of the sources which affect the harm they do. These are given in TABLE 2, and need some explanation. The following brief paragraphs indicate roughly the basis of the numerical comparisons made later.

TABLE 2. SOURCE CHARACTERISTICS

A.   Height, temperature, and volume of emissions
B.   Areal extent of source and consequent radiative effects.
C.   Concentration of emissions into periods of adverse weather.
D.   Closeness of source to site of damage
E.   Chemical (damaging) nature of pollutant

A. A source that emits pollution from tall chimneys at high temperature and in large volume from each chimney has its effluent diluted much more before it reaches the ground where the damage is done. The height and temperature of emission do not operate independently and must therefore be included in the same characteristic.

B. House smoke sources are usually spread over a large area. The consequence of this is that there is a loss of heat by radiation from the smoke which causes much greater stagnation of the air and a reduction in the dispersion away from the source region. A single large source such as a power station may produce a plume which covers quite a large total area, but because of its height it is usually extended by the wind which is almost invariably stronger than near the ground, and its radiative effects have a negligible effect on the motion of the air containing it.

C. If all pollution were emitted on windy days it would be carried away quickly enough for there to be no problem. Most of the stagnation of air occurs at night and in winter months, and it is at these times that domestic heating appliances are mainly used. Consequently any legislation to restrict their effluent would be more effective than similar legislation applied to trains which operate uniformly throughout the year and mainly by day. Restrictions on smoke emission would produce much less benefit in the case of trains operating on windy days, although they would cost as much to apply as on days of cold stagnant air.

D. A column of smoke from a chicken farm in the country does less damage than smoke from a house in a densely populated area. Rail traffic and power stations are much less concentrated into towns than road traffic and industry in general.

E. Smoke and SO_2 are more unpleasant and harmful than CO_2 and H_2O which are the main products from the use of fuel: in general large modern industries have a much cleaner and more efficient use of fuel, although some sources, particularly works processing large quantities of raw material (steel, brick, and chemical works for example) emit other pollutants.

Thirdly, in order to compare the relative "importance" of the classes of source in the community we need a numerical measure of their magnitude. This can be made on various bases, and is always open to criticism when a comparison is made between a house or power station, which is simply burning fuel, and other sources which are processing large quantities of material from which pollution emerges (as just indicated in E above). The measure used here is the total fuel consumption, but undoubtedly a more complex formula could be used which added a factor according to the amount of the pollution originating from the processed material.

It is important that a variation in one of the characteristics (A–E) should not have a direct influence on any of the others. The validity of the calculation depends on this. The factors in TABLE 3 are numerical estimates of the relative importance of the same characteristic for the different sources. They do not represent the relative importance

131

of the different characteristics for the same source: this ratio does not matter at all. The factors for a source are multiplied together to give the total factor, F for each source and it is only the ratios of the values of F for different sources that affect the answer. The values of one characteristic for all sources may be multiplied by the same number without affecting the result. Consequently we may take one of the sources and assign the factor 1 to it for all characteristics and choose the other factors accordingly. TABLE 3 gives estimates for Britain of 1952.

The fuel used (T) each year is given to the nearest 5 million tons. The distribution has changed very considerably since 1952 and the exact values are not important in the argument because the damage done per ton of fuel used varies by a much larger factor than the figures for any of the separate characteristics A to E. This is because sources which have a high figure for one characteristic tend to have high figures for others also. The total cost (G) of the damage and the damage (H) per ton of fuel used are determined from the assumption that the total annual damage done amounted to about £500 million. The values are rounded off to an accuracy of roughly 15 per cent.

TABLE 3. RELATIVE IMPORTANCE OF DIFFERENT KINDS OF AIR POLLUTION SOURCE

		Source						
	Characteristic	I Electric power	II Large industry	III Small industry	IV Offices, large buildings	V Houses	VI Road traffic	VII Trains, ships, planes
A	Heat, Temp.	1	3	5	10	30	40	30
B	Area and radiation	1	1·5	1·5	4	5	2	1
C	Weather	1	0·6	0·6	3	3	0·6	0·6
D	Closeness	1	1·5	1·5	2·5	3	6	1·5
E	Nastiness	1	2	4	1·3	5	2	3
F	Total	1	8	27	400	2300	580	80
T	Fuel used (10^6 tons yr^{-1})	35	30	25	10	25	10	20
G	Total damage (£ million)	0·3	2	5	30	400	40	10
H	Damage per ton (£)	0·01	0·5	2	3	16	4	6

In making use of this kind of estimate we have to appreciate that although the end result is only a rough quantification of what many observant people have said anyway, it does assign specific causes. New power stations cannot be detected by their pollution in urban areas because they contribute such a small fraction of the total pollution at ground level, whereas in towns with no smoke control whiffs of coal smoke can be detected by the nose in every street. This is mainly represented in the relative figures of row A, which are not easy to alter by legislation except in the case of industry which can be encouraged to use taller and fewer chimneys.

The only way in which row B can be altered is by reduction of the visible content of the pollution. Thus if smokiness is greatly reduced almost all these numbers can be reduced nearly to the value for large industry. This value represents the effect of hygroscopic particles in encouraging fog and mist formation.

The factors in row C can only be altered by changing the time of year at which pollution is made. Cheap rates for off peak electricity, night storage of heat, and pump storage of hydro-electric power has the effect of distributing the generation of electric power more uniformly, but has a negligible effect on pollution damage because the damage from power stations is so small anyway. Such devices are therefore justified by the economy of the generating process alone. "Meteorological control", which means the shutting down of sources when the weather is adverse, has long been considered but is scarcely ever practicable except at great cost.

The numbers in row D can be altered by siting pollution sources outside urban areas, and this was already under way with power stations in 1952, otherwise the figure D I would have been relatively larger. But the other D factors are almost impossible to alter.

In row E the cleaning up of the effluent by the removal of smoke is most important and is quite practicable.

The numbers underlined in TABLE 3 are those which it was quite practicable to reduce, and it is interesting to see what has been achieved by the Clean Air Act of 1956 and subsequent regulations. The Alkali Inspector has imposed stricter conditions on chimney heights in industry and this has tended to make A II and III in all new industrial installations much less than the prevailing average. At the same time architects have become more conscious of the need to have effective chimneys on large buildings and A IV has tended to decrease over the years. Local authorities are now better advised by the Ministry of Housing (Ministry for the Environment) on how to determine what chimney heights are acceptable on new buildings.

It is notable, however, that in the years prior to 1956 there was a definite tendency for domestic chimneys to be designed to be much smaller and ineffective in getting smoke away from the roof of a house than were the house chimneys of a century or two earlier. Old chimneys were probably made tall in order to get a good draught rather than to effect better dispersion of the smoke, but they did assist the dispersion.

As already mentioned, D I was already being reduced by building new power stations out of towns, and this has been carried on to a greater extent in the last decade.

But clearly the greatest improvement in air quality has been effected by stopping the emission of the more objectionable components of the effluent. Dark smoke was prohibited by the Clean Air Act from industry, and smoke control was made feasible for domestic premises. In areas where control has been introduced a very obvious improvement has resulted in air quality.

There was much argument at the time the Clean Air Act was being prepared about what should be done to reduce the effects of SO_2. It was considered impossible to reduce the amount emitted and so, in the event, nothing was done at all, except to insist on higher chimneys for the larger producers of it (factors A I and II). At oil refineries there has been a change towards fewer and larger chimneys, and a similar change would have been very beneficial in the brick industry which has changed its practices very little.

Technological advance has made possible a reduction of fine dust emissions, and the visible smoke from road traffic has been decreased although much still remains to be done in the reduction of diesel smoke. Trains have been converted from steam to diesel and electric power and their damage reduced to less than half, but planes have turned from petrol to kerosene and to engines designed to operate most efficiently at higher altitudes, with the result that the exhaust from planes on the ground is more objectionable and creates serious nuisance near airports.

The most significant effect of the changes has been due to the reduction of EV to around 1·3 in urban smoke control areas. If smoke control were introduced everywhere such a change would cause a reduction of the damage by domestic effluents from £400 million to £100 million a year, and half the 1952 pollution problem would have been solved!

The pattern of fuel consumption is changing quite rapidly. Power stations use much more fuel, which is beneficial because so little damage is done, but road traffic also uses more so that the harm of traffic exhaust is probably approaching £80 million and is comparable with what domestic pollution will cause if the whole country is subject to smoke control.

Another feature of the change is that, as the air becomes cleaner, amenity is more highly valued and damage to health is greatly reduced. For this reason traffic exhaust will become more strongly objected to because it is damaging cleaner air. Traffic exhaust is also a much worse problem for a variety of reasons, mainly meteorological, in sunny regions like Los Angeles, and the budget for that region would look very different from the one presented here.

One of the more interesting results of smoke control has been that the concentration of SO_2 has not increased as much as the domestic and other urban production of it. This is represented by the factors B, because an effect of getting rid of smoke (in addition to stopping the damage it does, mainly to health) is to reduce the stagnation of the air: the other pollutants, such as SO_2, are then more effectively dispersed. There are clearly two advantages in concentrating on smoke reduction and they act independently and are therefore included in different factors B and E.

Now that we are on the way to eliminating the worse forms of air pollution further improvements will inevitably have less beneficial effect and will therefore be slower. Bonfires, agricultural smoke, mainly from burning stubble after the harvest, and odours from industry are likely to become more important forms of air pollution in the future, and they are not associated with the consumption of fuel for the production of heat or power. But more important perhaps will be the change in emphasis that will be given to different forms of damage to our environment. Water pollution, noise, and desolation of land will be seen as having a higher priority than further improvements in air cleanliness, and air pollution studies will turn more towards the global effects of low concentrations of pollution on the world wide scale.

Finally, it must be remembered that studies of this kind could be made for much smaller areas than Britain, such as a region with a dominant form of industry. It could well reveal locally that a reduction in the obnoxious components of industrial effluent would produce the most beneficial effect.

The analysis, and the measures recommended as a result of it, make use of judgments which can only be made with confidence after considerable experience in studying air pollution problems, and they show, as we shall see later, that some laws which may seem good lead to expense out of all proportion to the benefit which results.

British air pollution legislation is an example of practising the art of the relatively easily possible. The starting point of the thinking has always been the present rather than a theoretical goal. The law has required everyone in due time to use the best practicable means to eliminate special forms of air pollution. In special cases where it is desired to emit highly dangerous chemicals strict controls have been imposed as a result of frightening damage. The malignant growths induced in lungs due to beryllium used in the manufacture of fluorescent tubes led to strict control inside the factory: unfortunately the general public was not as well protected in that case, and a few people living in the neighbourhood contracted the disease before emissions to the atmosphere were finally stopped.

This illustrates the important point that our legislation has tended to be built up as a result of unfortunate experiences, and it is of a type best suited to the improvement of a bad situation. Our philosophy must be extended so that we can avert disasters due to new technologies and gross expansion of old ones. The experience has enabled us to develop a planning system which requires that all new installations be approved in detail before they may begin operation, and it enables them to be required to achieve the standards of the best. We need now to examine what other approaches are possible, and whether progress has been made in other countries along more profitable lines.

6. LEGISLATION AND PROSECUTION OF OFFENDERS

In designing legislation we have to consider more than the scientific justification of a regulation. The introduction of smoke control areas in Britain meant changing a way of life. Pursuing the art of the possible the government left it to the local authorities to decide the rate of the change, and after 15 years some have still made no effective move. An important feature of the change is the inspection of every house to determine the magnitude of the subsidy to be paid to the householder to lighten the cost of the conversion to him. Smoke control areas have been introduced at a rate determined by the speed of the conversion to heating systems which do not make smoke.

In such circumstances the question of prosecuting offenders scarcely arose because the conversions were supervised and the sale of smoky fuel in the control areas prohibited. The selling of unauthorized fuel rather than the emission of smoke was then watched carefully.

In the U.S. the philosophy tends to be towards making undesirable practices illegal and waiting for the possibility of prosecution to compel people to stop them. It invites arguments questioning the constitutional validity of stopping people doing what they have been allowed to do since time immemorial. When those arguments fail, as they must in the long run, although the run may nevertheless be excessively long, polluters question the evidence that a crime has been committed at all. Not only is it impossible to set up a monitoring system which can prove that a certain level of pollution was exceeded anywhere other than at the site of an instrument, but if the instrument happens to be well placed to record excesses it is impossible to attribute the excess to a particular source with certainty. Consequently every case can be argued at length.

If a prosecution for an offence against air pollution laws is successful what is the appropriate penalty? If we accept this as part of the problem we are envisaging the deterrent power of penalties as the main force behind the law and there is no end to the evil social consequences: bribery, the imposition of frivolous or excessively punitive fines, unending costly appeals, establishment of unhelpful precedents, regular payment of fines by continuing offenders, the whole gamut of legal (and illegal) devices designed to protect the perogatives of the citizen and thwart the intention of the law. It is obviously preferable to avoid the issue of penalties altogether by a system of licensing approved installations. If the conditions of the licence, which are individual to the installation and are specified when planning permission is granted, are not complied with the operation of the installation is prohibited. If an unlicensed process is operated the legal consequences are essentially simpler than if an offence against a general prohibition has been committed, and such offences are likely to be rare if a good relationship between the Alkali Inspectorate and the local planning authority on the one hand and the citizen on the other is built up. In practice a licensing authority can provide a helpful advisory service which encourages co-operation, whereas if offences have to be observed and proved in court an antipathy is created.

The rarity of prosecutions for offences against air pollution laws in Britain is taken in some parts of America as evidence of the ineffectiveness of the laws. Alternatively it is regarded as proof that the British are more docile (and therefore less enterprising) by nature or nurture. In fact it is the direct and valuable result of the system.

The tax system, laws of contract and trade, property rights and laws of inheritance, the establishment of rights by prolonged practice, all create a system, an environment, in which the individual plans his life's operations. The expectation that the same basic rights will be preserved induces a powerful conservatism, and in America the constitution is employed in the long drawn out legal wrangles that accompany social changes, and behind the legal scenes all sorts of commercial pressures and other threats are used to preserve powers which give economic advantage. Any cost analysis must take into account the probable costs of litigation and insurance against its failure, which illustrates how the artificialities of the way society is organized can distort judgments about what is of value, and to whom.

In some states in the U.S. the state governors have been threatened by citizens with legal proceedings for their failure to institute air pollution laws for the protection of the citizens. The consequence is that all state legislatures are under pressure to institute some sort of air pollution laws. The yardstick in such circumstances is not the effectiveness of the law or its suitability for the state in question but the nature of the public pressures and the precedents set by other states. Public pressures are usually misplaced because they originate in scaring headlines sought out by lay-journalists, science correspondents, and ambitious politicians, often with strong commercial connections which we would regard as improper in Britain. The precedents set by other states are often promoted by those states on the two grounds that their neighbour's air crosses the border and that industry should not be less restricted in a neighbouring state and thereby enticed into it. Many of the precedents have been set as a result of ill-informed public pressures.

The scientists who advise about legislative needs do not get a satisfactory hearing in

U.S. largely because of a tradition that as much as possible should be left to private research corporations and consultants. The whole approach becomes involved in the atmosphere of litigation which dominates thinking about restrictive laws, and becomes partisan.

Wherever there is a law setting out technical requirements it becomes part of the job of the consultants to advise on how the law can best be complied with or otherwise evaded. There is pressure for the law to take very specific form, like building regulations, for then it is easier to enforce compliance. In the case of air pollution the precise purpose of a regulation is not easy to define, and so air quality criteria which will make the requirements specific are often sought. The idea of setting up standards arises not only so that the routine requirements can be met but also so that an offence can be clearly defined. Both of these needs demand that criteria should be defined officially, and we shall discuss their validity in Section 8. In anticipation of that discussion it is appropriate to say here that the criteria can only serve to complicate the artificial structure of air pollution control and make it more arbitrary and ineffective in actually cleaning the air. They become standards of permissible pollution in so far as they are capable of having any real effect, and they support the concept that there exists a right to pollute the air at least up to the level of objectionable nuisance.

In Britain the 1968 law gave the government power to put serious pressure on local authorities to institute smoke control which according to the 1956 act they were merely permitted to do. It is hoped that the threat of action by the government will induce the "laggard" local authorities to act. Slow as it may be, this is preferable to having national air pollution laws which are "enforced" by threat of prosecution. Many states in the U.S. are equally laggard, but federal legislation, which is now expected, is unlikely to be enforced in the laggard states, especially if the means of enforcement is through the courts and the form of the law makes it difficult to enforce.

7. BLANKET LEGISLATION

Most sources of pollution are tolerable a large part of the time in many places. Pollution becomes objectionable because of the weather of the moment and the nature of the locality and its pollution sources. Blanket legislation is a method of control which seeks to limit the offence by prohibiting a much wider activity which is believed by some people to include the offence.

The obvious example of a blanket law is the speed limit. It is illogical in the extreme, but read on! The areas where limits are automatically enforced are defined by the presence of street lighting, which if anything, contributes to making fast traffic safer. The limit is operative in all weathers and at all times of day for all types of vehicle and driver. The law is not obeyed, and the accidents, whose supposed likelihood in the absence of the law was the law's justification, scarcely ever occur. It is even doubtful whether any are really avoided as a result of the limit. New justifications for controlling speed by law can be argued as a result of the increased traffic densities and the improved standards of driver and pedestrian behaviour, induced by the control, but on the whole the law is not enforced because it is not enforceable except, as in some parts of the U.S., in a crude sort of way by fear of occasional exemplary punitive fines and compulsory attendance at degrading roadcraft classes.

If we are to seek an adult society we must expect the citizen to act with discretion.

One consequence is that the speed limit has become a technical regulation which can be employed as a device to prosecute drivers behaving dangerously but whose real offence is difficult to prove. It also has a very valuable sort of advisory effect: if you exceed it you are under very special obligation to ensure that you do not endanger or inconvenience others.

This last result—which enables any complainant to say "you offended against the regulation" when he is really complaining of manners which are bad by any standard, might be applied in air pollution but none of the other arguments which are valid for speed limits hold for air pollution. An excess of speed has no bad effect whatever as such whereas an emission of pollution is always objectionable and it is so much more important that pollution laws should be enforced. In the absence of any regulation pollution is offensive, speed *per se* is not because it is a universally practical convenience which can be easily avoided on any specific occasion on which it might be a nuisance. It is therefore not profitable to have a law controlling air pollution flouted as widely as the speed limit laws. The latter serve a valid purpose even though it is known that it is not practicable to enforce them.

An example of a blanket pollution law is that now operating in some eastern states of the U.S. completely prohibiting the use of fuels containing more than a certain percentage of sulphur. We shall see in the next section that it is very doubtful whether sulphur is worth legislating against anyway; but, on the supposition that it is, the most effective way would be to operate the prohibition only for low level sources of pollution in cities. The benefit obtained from reducing the SO_2 emissions from tall power station chimneys is negligible, but at the same time the cost of providing alternative fuels is large. There may be areas where sulphur free natural gas is plentiful where other social consequences of a change of fuel usage are not harmful; but in some coal mining areas the accelerated decline of a declining industry may be very serious socially: a law enforcing rapid change may therefore have a bad total result, and Lord Robens has often been described as "reactionary" when he has with considerable justification pleaded for not too hasty implementation of clean air legislation in coal mining areas. Blanket laws can be as ruthless and as inefficient as cost analysis in a situation where a way of life is being changed.

It is prudent to suspect that there is a profit motive behind the scenes in every blanket type law, probably located in the natural gas industry which most easily produces clean fuel. Alternatively the laws originate in political groups anxious to produce action on the pollution front and divert attention from other forms of pollution which they have an interest in allowing to continue. The existence of pressures of this kind can bring blanket laws into existence and prevent other more useful laws from being enacted.

Blanket laws concerning car exhaust are advocated in many quarters. There is the anti-lead-in-petrol lobby advocating the blanket prohibition of added lead on health grounds. The health hazard is widely alleged to be negligible although there has scarcely been time for the long term effects to be assessed. Nevertheless there are sound reasons for prohibiting added lead because it poisons the catalyst in the afterburners used to reduce the emissions of hydrocarbons, but on that basis there is no point in prohibiting lead unless the afterburners are compulsory. President Nixon attempted to introduce a tax on added lead; he attempted to float in the tax on a tide of emotion against lead as a biological poison. It may well be that the sole result of this

A.E.—B

138

tax would have been to raise money and that lead pollution would not have been reduced at all. This illustrates the danger that pollution could be (and probably is being) used as a support motive for exercises to which it is irrelevant, the reasons given publicly being the wrong ones!

8. THE CRITERION OF PUBLIC HEALTH

Public health has for a long time been the sphere in which a basis for air pollution standards has been sought. There are very great difficulties in using damage to plant life or structural and other materials instead, because our knowledge is so limited. Most damage to vegetation is the result of occasional large doses of pollution and we do not know much about its ability to withstand a long term low concentration, except that it is considerable. On the other hand materials do not suffer noticeably in the catastrophic incidents but seem to suffer a slow deterioration over a long period on account of the small concentrations of pollution. People seem to have much more diverse characteristics because we know more about them, and we have an impossible problem in setting up any criteria on the basis of human health because a dose that the majority can experience with no perceptible after effects can kill off a few chronic bronchitics and new born babies, especially if there is influenza about.

If we establish, as a standard not to be exceeded, a level of pollution that could do no observable harm to anyone it would be quite impossible to attain it everywhere in our society now. There are many unresolved issues: should we encourage those weak members of society who might suffer to keep out of the way of pollution and to stay indoors on cold evenings? Or must we ensure that there is no pollution anywhere which might harm them in case they might happen to wander into it? What durations of large concentrations ought we to permit, or should we seek a total prohibition? These issues are at the very heart of the practical administration of any law based on the criterion of public health.

On the whole it seems thoroughly undesirable to establish a permissible level of pollution, to exceed which is a crime, because to approach it most of the time, while remaining below it, could be very harmful yet legal. Alternatively, the limit could be set low enough to prevent that, but then it would be unnecessarily low most of the time. In order to take account of the variety of effects that are relevant the law would have to be ridiculously complicated—ridiculously because the complexity would be instituted for the sake of self-consistency and not because of its known real value.

There are two aspects of public health where we can argue constructively: the first is when a specific rare pollutant is discharged into the atmosphere. In such cases, which are few, it is practicable to require the emissions to be made so that no one can be harmed. The extra cost of a specially tall chimney to emit brown oxide of nitrogen is justifiable on the grounds that it is easy to ensure safety for everyone. If the effluent from all domestic heaters in a town were fed into a single tall chimney the pollution problem could be solved by making it very tall, but this is not possible with thousands; therefore the quality of the emissions must be improved.

The second aspect of pollution involving public health is cigarette smoking. The damage incurred is so large that it makes almost all other major forms of pollution seem sweet by comparison. Attempts to discover the effects of urban air pollution on health usually show that they are obscured by much greater effects of cigarette smoking. It is hypocrisy to bother at all about health aspects of pollution not already dealt with in our legislation without regarding cigarette smoking as a major disaster.

Each year about a million years of useful adult life are lost as a result of smoking. It is the same order of magnitude as the time spent in smoking: each cigarette curtails a human life by about the time required to smoke it. There are about 100,000 deaths a year attributable to smoking diseases—lung cancer, chronic bronchitis, heart disease, stomach and bladder cancers and various others. In addition to deaths occurring prematurely (about 10 yr each) the later years of the life are spoiled by a variety of physical debilities—shortage of breath, loss of taste, deterioration of sight and smell, the disadvantage of addiction and dependence, pain in the later stages of disease, the squalor of the stench of smoking experienced by others.

The annual cost of smoking in lost employment in the United Kingdom is therefore of the order of £1000 million, and the cost in health services provided to patients equal to a large fraction of this. The wickedness of cigarette advertising is blatant. It seeks to hook the young to the addiction by misleading and incorrect implications about the social and physical qualities of smokers. It is tolerated only because it has grown up within a legal system which permitted it before the ill effects of cigarette smoking were known, and no other health aspect of air pollution deserves attention while society suffers this scourge. The situation is quite ridiculous because the government could make cigarette advertising and smoking in public places illegal without much opposition (though it is thought without much thanks in the form of votes). The damage done by each cigarette is not to be compared with the tax collected because the tax could easily be collected in other ways without making that wretched section of the community, the smokers, more wretched by discriminating taxation. In so far as such a high tax as exists on tobacco is justified because tobacco is harmful it provides a good example of how disastrous pollution cannot be reduced significantly by fiscal means as long as the polluter can choose to pollute and the source can be promoted by advertising.

If we were to legalize cannabis, as some people recommend, it would be necessary to consider whether "pot-advertising" and "pot-promotion" should be permitted and whether the drug should be taxed. On the basis of our experience with cigarette smoking the answer must certainly be "no" in both cases. This does not mean that cannabis should not be legalized but that its promotion should be recognized as highly obscene and liable to deprave and be censored accordingly.

In summary, therefore, now that the means to abolish smoke has been provided and is known to be effective, health does not at present provide us with criteria for future steps in ordinary air pollution legislation.

This conclusion may not endure. It required a 30-yr experiment on humanity to discover the harmful effects of smoking and a disaster beside which the London smog of 1952 is trivial, now afflicts us. There is little evidence yet that even in Los Angeles the sunshine smog resulting from car exhaust produces any irreparable ill effects. Unpleasant though the sunshine smog incidents are, they may be no worse than minor scratches and bruises from which we are well equipped to recover—but we do not yet know this for certain. In the meantime our only protection is to place a greater value on the amentity which car exhaust tends to destroy. Above all we must not use reasoning which carries emotional weight with the uninformed public but which is known to be incorrect.

In this category comes the treatment of SO_2, CO, and lead in petrol. It is known that sulphur dioxide causes a rasping choking cough if inhaled in large concentrations. It is also a very plentiful pollutant because it occurs in most fuel. Since it is fairly easy to measure it has become an indicator of pollution levels since the burning of fuel is the main source of pollution. Yet it is not incriminated as a major cause of damage when the concentration is low: its main harm is the result of it being absorbed on the surface of solid pollution particles such as smoke, and when the smoke is removed the SO_2 is acceptable, being invisible and in usual concentrations unsmellable. Moreover, the atmosphere is the source of sulphur on which almost all vegetation depends for its healthy existence.

Carbon monoxide is the component of car exhaust which kills those who commit suicide by shutting themselves inside the garage with the car engine running. Death occurs when it is absorbed to such an extent that the blood is prevented from carrying enough oxygen to the brain. It has no measurable effect in small doses and has no odour or colour. In a dense traffic jam the levels reached in the blood are less than half those reached in cigarette smoking, and about the same as those reached by holding conversation with a smoker. The CO absorbed is breathed out again in fresh air and no permanent damage is done. It may even be that smokers experience some sedative action, which they like, from the CO they inhale.

Consequently, we ought to concentrate our attention on the hydrocarbons and oxides of nitrogen in car exhaust (the continuous emission of visible smoke can easily be prevented and should be made illegal). We have to beware because most practicable methods of reducing the amount of unburnt hydrocarbon increase the amount of oxides of nitrogen, and we do not yet know their relative importance as polluters. Certainly the photochemical effects prevalent in Los Angeles and which are occasionally observed in other places are scarcely ever possible in Britain because of differences of climate. Therefore measures instituted in California may be quite pointless elsewhere. Furthermore, we may need to wait for the maturation of the long term experiment on the population of Los Angeles and other big cities before we know whether we are entitled to complain about car exhaust on health grounds.

On the whole odours are not damaging to health but only to amenity. They operate in a very complicated way and the concept of a threshold concentration below which they are not detectable is misleading. They are detectable mainly when first sniffed after breathing fresh air. If breathed continuously in fairly high concentrations they cease to be noticeable. Their source is difficult to locate, and at oil refineries is commonly a leak which it has not been possible to find. On the other hand it is likely to become easier to detect them in the more distant future, and then it will be a question of eliminating them at source where possible. The immediate prospect is, unfortunately, that the oil industry will continue to be surrounded by unpleasant odours but without harm to health.

9. AIR QUALITY STANDARDS

The theoretical purpose of an air quality criterion is to provide a standard by which a community can decide whether its air is acceptably clean or not. Such a standard must necessarily be arbitrary because it must be based on a criterion (such as the number of deaths produced per million of the population) which is itself arbitrary, and determined subjectively. Even supposing that a number of acceptable deaths attributable to pollution were agreed this cannot be converted into a threshold pollution concentration because the necessary medical evidence does not exist. When asked to provide a numerical estimate based on such data as is known, medical men almost always come out with a new horror story about smoking, and say that all other effects on health are obscured by its dominant effect.

We have already seen that pollutants act synergistically (the SO_2 does much more harm in the presence of smoke) so that their effects are not additive, and acceptable concentrations ought therefore to be expressed as functions of the concentrations of other pollutants. This is not a practicable possibility in regulations.

If we suppose that an acceptable level were defined we would still have the insoluble problem of converting this into what would be acceptable in terms of the thousandfold variation in concentration observed from day to day, and often from minute to minute when the concentrations are high, because of the variations of the wind and weather.

Even supposing that that problem were also solved, we would still have to estimate what concentrations would be caused by a theoretical specified source of pollution. This cannot be done for typical weather conditions because it is too complex a problem. It can be done roughly for certain special conditions, but they are not usually the conditions which cause pollution problems, and we do not know the frequency of occurrence of these and other weather types, and cannot forecast them because of lack of observations from the past.

If it were possible to predict the pollution to be produced by a specified source in a specified climate, it would still not be possible to identify a particular source for certain as responsible for an incident in which pollution was measured to have risen above the standard, because there is no accurate enough routine information about the wind. Even if all possible information about the wind were available in a perfect world, it would be necessary to have universal measurement of pollution in order to discover whether the standard were being maintained.

Action to be taken through the courts as a result of an alleged infringement of the air quality standard is now beyond human comprehension because there are at least six stages at which the case could be argued, and probably faulted for uncertainty at the technical level. The constitutional problems of attempting to set up legal standards in such a situation are a lawyer's gold mine. It is no cause for surprise that in the U.S. almost every argument about methods of ensuring clean air becomes quickly a discussion of how the legal aspects should be handled. In the political atmosphere there prevailing every scientific uncertainty begets a greater legal problem which absorbs more intellectual effort than the scientific one!

By contrast there is a perfectly sensible procedure which can be adopted without involving any of the uncertain issues just discussed. It is to gain a fair knowledge of what the present practice is and investigate how much we can improve upon the situation without excessive cost. This is in fact what has been done in Britain: having decided that existing levels of pollution were undesirable, means within our reach to improve the air quality were investigated and the most effective and practical ones were chosen. It was, perhaps, a matter of luck that a practicable programme which could produce useful results was possible, and it is likewise a matter of misfortune for California that there is no easy solution to the sunshine smog problem in an automobile civilization.

In determining chimney heights the basis of British procedure is to obtain a fair idea of current practice and accumulate experience of what has been found acceptable: all new installations are then required at least to meet the best standards and so the general standard is steadily improved.

These criteria are all right in a country that is already very fully developed and where the experience exists; but where neither circumstance prevails the implied advice is that the experience should be sought elsewhere. The alternative is to develop theoretical models such as the "airshed" idea, which attempts to estimate what burden of pollution may safely be emitted into the air of a region, certain suppositions being made about the air movement. The main difficulty in that method is that our knowledge of upward diffusion is almost non-existent, and such knowledge as we have is based on diffusion models applied to ground level pollution produced by elevated sources, and the values measured are so scattered that almost any reasonable model can be fitted. Nothing quantitative is really known, except what can be deduced from visual observations, about the transport of pollution to points in the atmosphere higher than the source, and such knowledge is a *sine qua non* for any reliable model.

Along with the definition of air quality standards goes the idea of "air resource management". There are times when one's basic philosophy initiates a rebellion against the very conception of an idea: for me this is such an idea. It must be admitted that before long global pollution problems will become our major concern and in that sense our husbandry of resources will have to extend to the whole atmosphere. Nevertheless I think it is part of the legacy of the industrial revolution which we must throw off to assume and even hope that man will control the world. Till now the magnitude of the world has scarcely been touched by industrial man's activities, but at last we have to draw back in awe from the limit and deliberately leave our successors room for manoeuvre. By pressing our occupation and development of the world to the limit we would reduce the variety and interest in living in it. The forests, deserts, and snowy wastes provide a vast reservoir of emptiness in which our air can recover its natural purity and through which we can travel and fill our minds with awe of nature instead of with the oppressive proximity of our own multitudes. Science stops short of trying to understand nothing, but technology and industry must gradually reduce their material turnover to a level at which the amount of spoiled land and polluted sea and air increases no more.

Therefore in the end we must plan to manage only a small part of the air and take only a little of it as our resource.

10. GLOBAL POLLUTION

Although we learn to avoid high concentrations of pollution near the sources and in cities, we may nevertheless increase the total output in such a way as to alter the composition of the atmosphere. Two important components have been considered ever since causes for the ice ages were sought; these are the CO_2 content and the turbidity (fine "dust" content) of the atmosphere, both of which have varied greatly in geological time. The CO_2 content has increased by about 20 per cent in this century and is increasing at the present time by about 0·7 per cent per year. At the same time the mean annual temperature of the air appears also to have risen about 0·5°C. This temperature rise is a very crude estimate, and the rise in Britain has been about twice that amount and in Franz Josef land about five times as much. The rise is attributed by some writers to the greenhouse effect, whereby the extra CO_2 reduces the loss of heat by radiation from the lower layers of the atmosphere. An alternative explanation may be found in sunspot activity which is thought to be an indicator of total radiation from the sun. Sunspot changes have been observed for some time. They have a cycle of approximately 11 yr superposed on other longer term changes for which there is no explanation supported by evidence. The 11-yr cycle is possibly caused by tides due to Saturn and Jupiter, or by their capture of interstellar gas in the track of the sun so that the sun's tunnelling shows these variations.

In the last 20 yr there has been a fall in temperature which is very pronounced in the Arctic Ocean, and much more marked in Britain than in the world as a whole. This could be due to a change in the density of the interstellar gas through which the sun is tunnelling, or to an increase in atmospheric turbidity, which has definitely been observed. It cannot be connected with CO_2 which continues to increase.

The turbidity increase is probably due to increased cultivation, mainly in the tropics, accompanied by burning of agricultural refuse and debris from clearing of forests. It may be due to an increase in material carried by the atmosphere from desert regions. Whatever the cause, the decrease in temperature is more rapid than the rise of the previous 70 yr. Volcanic dust has long been associated with cold eras of the past by many investigators of climatic change.

Since these changes in temperature are produced as much by changes in the circulation of the atmosphere as by the radiation balance it is possible that effects may possess a sort of instability within the atmosphere, and once the arctic sea begins to freeze more extensively, the mechanisms which began it may be accelerated for a time. Because there are at present no theories or calculable models which can investigate quantitatively all the possible causes, certainly not simultaneously, there is bound to be a great deal of uncertainty in this field for some time. Many changes could be brought about by internal causes, such as ocean freezing, as important as those due to cooling of tropical air by increased smoke. None of the recent changes is as large as those that have occurred, on account of unknown causes, in historic times. Such changes have been indicated by the positions of glaciers in Europe, and variations in climate in Greenland and Alaska.

The world consumption of fuel and production of smoke and dust are both likely to rise until the end of this century before levelling off, as they ultimately must, when the world population is stabilized, and so this is an important, but very speculative realm of study in which vigilance is important.

On to this scene has come the SST controversy. According to the opponents of supersonic transport planes the increased water vapour emitted into the stratosphere will alter the radiation balance and cause a warming similar to that produced by a CO_2 increase. Such calculations as have been made, based on the assumptions (1) that 400 SST's would emit 150,000 tons of water vapour into the stratosphere daily, (2) that the residence time of gas particles in the stratosphere is about 10 yr and that the equilibrium water vapour content of the stratosphere would therefore rise to about twice its present value, suggest that as a result the air temperature at sea level would rise by about 0·6°C.

There are several uncertainties in this argument. The average residence time in the stratosphere is almost certainly only a small fraction of the 10 years supposed in the above calculation, and as a consequence the equilibrium water vapour content would be much less than twice the present value. It would be achieved after about a decade of flying with the full 400 aircraft, and the new temperature would be attained after another decade or two. The effect would depend on whether the whole atmosphere or one part preferentially were warmed because the gradients determine the vigour of the circulations which in turn determine the temperature changes. At the same time as any change was being produced other changes would undoubtedly be occurring due to changes in the sun's radiation, the CO_2 content, and the turbidity, and perhaps other causes such as irrigation of deserts.

Another suggestion made recently by H. Johnston is that by a catalytic action a component of aircraft exhaust might cause a depletion of the ozone in the layers flown in. Oxides of nitrogen are invoked as the catalyst and if the ozone were destroyed it is argued that harmful u.v. radiation would reach the ground. The argument can be faulted for several quite independent reasons.

Supposing that the ozone in the layers flown in, namely from the tropopause up to 20 and occasionally perhaps as far as 25 km above sea level were removed there would be no significant increase in the u.v. radiation reaching the ground because all that is absorbed is absorbed higher up.

There is already a considerable amount of oxides of nitrogen reaching these layers by the natural mixing of air across the tropopause in storms and by occasional penetration after volcanic eruptions (and nuclear explosions). In effect, therefore, nature has already performed an experiment and shown that the result would not be to deplete the ozone.

In the absorbing layers from 30 up to a little over 50 km the absorption produces warming which is the cause of the very stable stratification in those layers. It is also accompanied by the production of ozone, and by its decomposition at approximately the same rate, to maintain the absorption. The introduction of catalysts to introduce a new mechanism for decomposition would have little effect at the concentrations that could possibly be caused by aircraft.

Even if aircraft could introduce catalyst which did have an effect, the mixing processes are very slow and only a small fraction of the air would be affected. If the mixing were rapid enough to distribute the pollutant throughout the whole layer it would also mix it out of the layer into the layers above and below which contain oxides of nitrogen naturally—those below for reasons already stated, those above on account of the impact of meteorites (shooting stars).

The proposers of the theory have considered themselves immune from attack on the grounds that their proposed catalytic reactions might occur and that we ought not to risk the occurrence until we are certain about them: they know too that experiments at the low pressures in question are almost impossible to conduct in a laboratory on the ground because of the large volume that would be necessary in order to avoid wall effects. But nature has already proved that if there is any effect it is small and has always been operating.

This type of theory is of particular interest because it is in the category of science fiction: it predicts a catastrophic effect from some really rather small activity by man and is unconcerned with the damage done by pollution at ground level when large tonnages of pollution are emitted into rather stagnant air or water. It is "instant science" and is followed by "instant morals" and "instant politics". It is born of a naive view of nature which while pretending to abhor the dreadful effects man's activities may have nevertheless believes that man has it within his power by subtle means to alter the enormous complexity and infinite variety of nature. In fact nature has already done all possible rainmaking and similar experiments, and is continually doing them and has arrived at an equilibrium which is not in the least unstable. The system is not waiting to be tipped out of a fine balance; it suffers continual internal turmoil far greater than man creates. The true student of nature knows this, and is concerned with the continual expansion of polluted areas around our centres of dense population where there is interference in the ecology by very large tonnages of waste products.

Of all the causes of change, the SST's alone would be fairly easy to stop, for they could easily be replaced in 20 yr by jet aircraft flying at present day altitudes. They would be under the control of three or at most four countries, whereas the other causes of change for which man might be responsible would involve a large fraction of humanity and their whole way of life. In this sense the SST's would give us the only "controlled experiment" in climatic change that we are likely to be able to conduct, and could be commended for that reason. Far from creating further uncertainties it is likely (though only if the effect is at least as large as the gloomiest prediction) to be one of the few safe sources of firm knowledge on the subject. The pollution in question is of the same kind as from the rest of the combustion of oil in the world and poses no particular problem in that respect.

11. BACKGROUND POLLUTION

Even if the effects of pollution are reduced to acceptable levels near the source there may be adverse effects due to the continual presence of a background of pollution caused by distant sources up to 1000 km or so away. If there are such effects they must be due to special local problems at the receiving end. An example of this is suspected, but not established for certain, in Norway and Sweden, where the soil is very shallow and contains no lime. Consequently rain which is acidified by SO_2 may not be neutralized before it enters rivers and lakes and may have a harmful effect on fish there. Although the effect is not in doubt, the origin of the SO_2 in question is uncertain. There are several strong sources in Scandinavia such as towns like Gothenberg and Copenhagen and these are the most likely source. The North Sea emits about as much sulphur in the form of H_2S which is soon oxidized as Britain does in the form of SO_2, on account of the decaying vegetation, and is probably as important a source of SO_2 as the industrial areas of Germany and Britain combined, as far as Scandinavia is concerned. The North Sea is a strong source largely on account of the polluted rivers flowing into it from the densely populated areas of N.W. Europe, but the point at issue in the argument is whether the British policy of dispersing power station effluent from tall stacks is a cause of such problem as exists in Sweden and Norway due to sulphur in rain water.

A very rough estimate indicates that probably much less than about 1 per cent of Britain's emitted sulphur is deposited on the whole of Sweden, whereas much more is both absorbed directly by the vegetation of the smaller area of Britain, and deposited on Britain by rain with no damaging effect at all because of the abundance of lime in the soil. In this connection it must be remembered that the world's vegetation depends on the atmosphere for most of the sulphur it requires for growth, and that SO_2

should not necessarily be regarded as pollution. The seas of the world emit as much sulphur as the world's industrial areas.

The main objection to background pollution is aesthetic. The loss of amenity can be very real when the visibility is much reduced, but even within Britain the damaging effect of pollution is confined to areas close to dense population and occurs close to the source.

There can, however, be said to be a background pollution problem within these areas, even when the pollution within a mile of the source has been reduced to acceptable levels. In industrial areas surrounded by mountains a serious smog problem can occur in calm weather in winter, and this is very serious in N.W. Bohemia where the coal used is 4 per cent sulphur and the number of days of winter calms may be as high as 35 in a season.

When seen from the air the densely populated regions of the Western world do appear to be highly polluted, but the visibility over the Eastern Mediterranean or over New Delhi on account of dust (and smoke?) from the desert regions of the world can be as bad as over Pennsylvania, Derbyshire, or the Ruhr.

It does appear, therefore, that the background pollution problem is serious (1) in regions shut in by mountains, (2) as a possible cause of world climatic change, or (3) as a deterioration of amenity, but not as pollution causing direct damage.

12. AIR, WATER AND LAND POLLUTION

While water pollution in rivers and small lakes is easier to remove or neutralize because of the confined nature of the water mass, the harmful effects are within the water itself and are usually more serious than those of air pollution. Not only are they worse, but restoration of the unpolluted state is more difficult, and takes years, or perhaps even decades after the pollution source is stopped. Thus Lake Erie and the Baltic Sea may remain seriously damaged beyond the end of this century even if the pollution is controlled by 1980.

Land spoilation is only rectified by nature in centuries. By contrast the air is cleaned in a matter of a very few days in temperate latitudes and very few weeks in the tropics. Air pollution is dispersed rapidly in most cases up to the height of the cloud base, and then requires the ascent of air above that level to form rain-bearing clouds before it is washed out. In drizzly wet smog situations the residence time of pollution particles may be of the order of a day only.

The consequence of the shortness of the healing time of air damage is that improvements are immediate when the source of pollution is controlled. Except for the problems discussed in the next section, the task of control in Britain is really rather obvious for we know how to eliminate smoke pollution almost completely, and that causes most of the damage.

In the case of water and land pollution there are property rights and a dreadful inheritance of bad practices, precedents and accumulated damage to make the situation much more serious. Consequently we must be extremely wary of solving air pollution problems by wet-scrubbing of effluent gases, because a more serious water pollution problem could easily be caused. While our air quality in Britain is being improved, our supplies of clean water are becoming inadequate, and our planning needs a bigger investment in clean water than in clean air.

13. OUTSTANDING PROBLEMS OF AIR POLLUTION

There are three problems of air pollution in Britain which stand out among the others. The first is the need to plan the ultimate population density in the country beyond which a deterioration in the quality of life would be unavoidable. The air pollution aspect of this is to determine the total energy production (fuel consumption) density. In modern industry we have to plan for several years ahead, so great is the investment. It is quite obvious that crude extrapolations of present trends, whether they are assumed to be linear or exponential, are not a very intelligent exercise. In the case of many products there is obviously a limit, which may be reached fairly soon after their first introduction, when demand levels off because it is satiated. The disagreements among expert economists on almost all predictions arise because they see the exercise primarily as one of prediction, not of planning. The result is often that gigantic promotion exercises have to be undertaken in order to create the predicted demand. We have come in some cases, to the ludicrous extreme in which, because it has been discovered that a demand can be created, it has been decided to create a particular demand to suit the producer. The Ford Mustang and Capri cars are instances of this. A taste for the Capri design has been artificially fostered, even though it represents the satisfaction of no pre-existing need. I shall refer to this case later.

It is the task of intelligent government to study the human needs that will have to be met in a "fully occupied" territory in order to determine the way in which an economy, which has evolved so that it is healthy only when it has a rapid increase in production, can be transformed into one with a more or less static population, static energy consumption, and static material turnover, but with a continual evolution of style and purpose. In such a community, there must be waste land, wilderness and unused sites, all the time, or the planning cannot evolve: therefore we must plan to avoid maximizing the use of resources at any time. In this technological age evolution is the stuff of life and so it must have elbow room to facilitate the changes. It is imperative that it should not be allowed to expand to the full with excessive capital investment and land utilization nor evolve laws or accounting systems which prejudice judgments as to what is worth while. Nor should it allow incentives to operate which direct effort into anti-social activities. How this is to be done is not the present subject of discusion: my purpose is simply to emphasize that pollution becomes a serious problem when these issues begin to become important: it is an indicator that the moment has arrived to rethink our objectives and the motivation of the economy. We can say that we should seek to increase the real value of every ton of goods produced, in order to increase our wealth without consuming more material, and by the technological advance which this implies actually use less material in our daily life.

The second major problem is that of road traffic. It is, of course, a problem as traffic, but its noise and its stink damage amenities either directly or because of the formation of sunshine smog. Traffic densities have already reached the maximum possible in many parts of many large cities. The prospect for the future is that the area in which the maximum is reached will expand. In planning our cities we have to recognize that the exhaust is mainly diffused upwards in Britain: consequently, as in the case of domestic smoke, the greater part of the fumes which are objectionable at any point have been emitted within a hundred yards or so of it. Background pollution is not normally a problem, although it is occasionally objectionable in the form of accumulated local pollution on very calm days. The only possible remedy is by improving the quality of the exhaust itself, and this will have to be done unless there is a radical change in engine design to one with a naturally cleaner exhaust.

The pressing question is whether this should be done by blanket legislation or whether some sort of local licensing should be evolved so that the extra expense of an engine with improved exhaust will only have to be paid in places, such as Los Angeles, where an improvement is a must. One possibility would be gradually to establish clean exhaust zones into which only licensed cars, with regularly tested clean exhausts, could enter. Like smoke control areas these would be extended in time, and this would be a natural way to get rid of a nuisance which has already developed to the stage where it is urgent to ameliorate existing conditions. Blanket legislation under which all cars were compelled to satisfy the clean exhaust code from a certain date would be more costly, and a less effective use of resources. A second advantage of the development of clean exhaust zones is that the methods of cleaning the exhaust could evolve as a result of a growing experience. To compel all cars to comply from the beginning of the operation with the code would mean installing large numbers of gadgets whose effect had never been tested *en masse*. If the requirement were imposed only on new cars, the maximum benefit would not be attained because old cars make worse exhaust. The best result would be gained by making clean exhaust a requirement in the areas where the traffic most often reaches maximum density, because no exhaust will be quite clean, and so the reduction ought first to be made where the concentration of pollution is greatest.

Certain improvements could be made by blanket legislation because they do not require the modification of most existing cars. Thus the emission of visible smoke in normal running could be made an offence: smoke tickets could be issued like parking tickets with an additional penalty that the vehicle licence would have to be surrendered until the vehicle had been made smokeless. Smoky exhausts are undoubtedly the worst offenders, and because they produce an objectionable level of pollution on their own they are worth eliminating by blanket legislation.

In planning the control of car exhaust we have to have a credible estimate of vehicle numbers, not a ludicrous extrapolation of recent trends. We need, perhaps, to place more emphasis on vehicle design, for too many vehicles are too big. If the majority of families with a commuter parent are likely to acquire two cars, a commuter car should be a subject of serious design. The conventional 4–5 seater with large luggage compartment consumes more fuel and therefore makes more pollution than is justified for solo commuting. The lesson of the Ford Capri, which occupies more road space for less carrying capacity than the Escort or Cortina and for much less than the BMC 1100–1300 range is that products ought not to be allowed to be promoted on the basis of sales techniques which are irrelevant to social needs.

We have undoubtedly been reluctant to have bureaucratic tribunals or committees sitting in judgment upon designs because so often design, especially when it is for its own sake, is utterly arbitrary and cannot be censored fairly. But now we have social needs emerging which we ought to take a lot of trouble to codify. In so doing we would hope to promote the best ideas and issue challenges to designers which they would take up.

The same problem is in the course of being overcome in building design. Architects have only lately begun to recognize the need for the effective dispersion of the pollution from buildings: they have long regarded chimneys as undesirable, and therefore to be hidden, but are now beginning to think in new terms which represent social purposes rather than competitive design gimmickry. They are always compelled to act this way when space is at a premium, and they must now do it when pollution must be dispersed.

In the case of vehicles a fashion could be evolved in a rational society in which potentially obsolescent body design would be a factor tending to reduce rather than promote sales. Vehicles and their design should be expected to last for one or two decades or more as the Mini and Volkswagen have done. Certain features could be obligatory such as vertical exhausts on all tractors, refuse vehicles, articulated vehicles, and certain types of truck, van, and bus, because emission near to the ground increases the concentrations within streets.

In making these points I am not selecting permanent priorities, but indicating that in a polluted society our idea of legality and legitimate promotional activity must be changed. We may have to change our criteria of obscenity so that promotion of sales for the sake primarily of profits rather than to fill a social need should be treated as obscene.

Enticement to acquisitive gluttony and the advertisement of any extravagance that offends the senses of other citizens by smell, appearance, noise, or the occupation of public space should be censored.

Such new criteria must be borne in mind in considering developments in research or commerce. As we become more permissive in many matters of personal conduct, so we may tighten our morality in respect of property and greed so as to make social outcasts of those who possess or desire excessive pollution-making wealth.

The outcome of these considerations is that it is not really enough to react with indignation, enthusiasm, or some other immediate emotion to offensive pollution, as the public have been encouraged to do over carbon monoxide and lead in petrol. We have to eliminate the offence according to a legitimate scale of priorities and examine how our social values must change so as to reduce the making of pollution. After all pollution is an unpleasant consequence of excess of a way of life that was acceptable in earlier circumstances, and when such a way of life is changed we are glad.

The third great problem of air pollution is to develop the practitioner's art on a wider scale. This art is already practised by the Alkali Inspectors. They are the arbiters appointed by law to decide, arbitrarily, what polluting installations shall be permitted. They are successful because they are more expert in the matter than the great majority of those they control; from the expert few they are quick to learn. The fact that they operate almost without taking any legal proceedings, even though they have the power to do so, indicates that they represent a much more acceptable system than one in which a law is promulgated for anyone to break and face prosecution if they dare. The reason for this success is obvious: they match the requirements they impose to the needs of the case and the means available, so that they are always seen to be reasonable. A specific law imposed according to its letter without discrimination as between cases is often obviously foolish and does not represent a sensible scheme of priorities.

This kind of practitioner is needed because every situation has its own peculiarities which are dominant. It is only when a respected public servant has the power and discretion to decide that these peculiarities can be taken into account in formulating the requirements to be met before a licence to operate is issued. The features peculiar to the case may include the nature of the district, its planned commercial evolution, its characteristic weather, and the economic needs of the industry.

At the present time many local authorities' Public Health Inspectors are rapidly developing the practitioner's art in relation to pollution due to sources outside the jurisdiction of the Alkali Inspector, but planning authorities do not at present consider air pollution in their promotion of new towns and suburbs. Clearly practitioners in this field are required as much as to deal with problems of drainage, transport, and the provision of sites for public buildings such as pubs, garages, shops, churches, and other meeting places. These practitioners could handle noise problems also.

The practitioner's knowledge is required also by many research scientists and engineers. Anyone planning a new industry, a new process, or making a field investigation (as opposed to a laboratory one) needs it. This is particularly true of laboratory scientists whose conception of the messy complex world of reality outside is often utterly naive—as naive almost as the uninformed politician's picture of science and "the scientific mind". The knowledge is needed above all in order to discover the priorities of the kind indicated in this essay: it is acquired by practice.

14. GROWTH AND THE ENVIRONMENT

In his Fabian Society tract *A social democratic Britain* Anthony Crosland attacks the environmentalists: "Their approach is hostile to growth in principle and indifferent to the needs of ordinary people. It has a manifest class bias . . . they are little concerned with the far more desperate problem of the urban environment in which 80 per cent of our fellow citizens live". Ordinary people are entitled to the products of advanced technology flaunted by the wealthy and the real issue is not the preservation of the environment for the few but how, and even whether it can be preserved at all.

We have to accept that tigers and other noble creatures are not now able to preserve themselves because man is occupying all the living space. Tigers are a much more valuable inheritance that we can pass on to posterity than great works of art because they cannot possibly be recreated once they become extinct. Equally the Australian Aborigines may become extinct, together with many other human societies which are being squeezed out by the technological and population expansions. Crosland has not answered this problem at all because he writes in terms of Britain's problems, not world ones.

When we accept that all citizens are entitled to the benefits of modern technology we cannot be indiscriminate about this. There will always be inventions which at any moment only a minority of humanity can possess, and it is important to prevent growth in industries which only make a contribution to living by geographical expansion, and use up more space. The longer we go on developing along the same old lines the problem will get worse. As Crosland says: "It is largely a backlog problem— the legacy of 100 years of unplanned growth. It is a problem of *existing* slum housing, polluted rivers, derelict land, and belching factories. Even if we stopped all further growth tomorrow we should still need to spend huge additional sums on coping with pollution". Which means that we must not continue in the old way, or problems will get worse. Of course I quite agree that such growth as takes place ought to be planned in the public interest; and there are many scientists, who are "environmentalists", who are not interested in any sort of privilege, but who see that the political objectives in growth are not enough to determine the course it should take. We cannot assume that every new convenience of modern technology should be mass produced for the use of the common man as soon as it is invented (or at least when its potential and immediate benefits become widely known). For example the Zürichersee fisheries were quickly but unexpectedly destroyed when the towns around the lake were fitted with flush toilets, and it is not in the least obvious which of the inventions can safely be made available to everyone. It is also clear that the inventive effort should go into technologies which produce products of the largest possible value per ton, and which raise our standard of living without increasing our output of waste.

But even that is not an adequate policy for humanity. If the present standard of living of the better-off half of New York's population were enjoyed by the whole, the destruction of the water of New York Harbour as an amenity would progress so rapidly that it would be dangerous within 5 yr. Even now it is scarcely safe to bathe in the East River. If that standard of living were enjoyed by all Indians the Ganges would be so polluted that life in its valley could not continue that way. In fact the total water supply there is at present incapable of supporting anywhere near that standard of living, if that standard be measured by material consumption.

Just as Crosland complains that the aristocratic gentlemen of Britain are trying to keep the best of the countryside for themselves, to say that growth is a must for Britain is to make out of our whole nation a privileged minority whose standards cannot possibly be approached by the majority of humanity.

Thus, however urgent economic growth may seem to be, undiscriminating growth will actually worsen our problems. When Crosland says "So the case for growth remains unshaken" he has not answered this point. I am not accusing him of being unaware of it because he does make his qualifications. But this issue does not arise out of political considerations; it arises because we are filling the world to capacity, and then political changes are imperative in order to devise a world fit to live in when growth of tonnage consumed can no longer increase—and indeed must decrease.

These issues ought not to be avoided because of the urgency of immediate economic problems. As we plan to live with less pollution we must also plan to slow down growth, and the fact that we think we need growth now must not obscure the fact of the physical limitations of the world.

Perhaps we can leave the problem of how to live together in a full world until later, but at the technological level the task is urgent now: to develop processes which give greater benefits with a smaller material turnover and waste. Politically we shall need physical elbow room, and the danger of too rapid growth is that we shall soon lose all choice in our planning through lack of space for new developments.

REFERENCES

Since this essay is essentially a discussion of method and attitude, it is not designed to be a source paper for other people's research findings. The following sources are recommended as being more specially concerned with the quantitative estimates used to illustrate the arguments. In some cases approximations have been made to simplify the argument in this paper, and such rough estimates should not be quoted simply on authority but with appreciation of the arguments which led to them.

ASHBY SIR E. (1971) *Royal Commission on Environmental Pollution—First Report*. HMSO Cmnd 4585.

LAWTHER P. J. and COMMINS B. T. (1970) Cigarette smoking and exposure to carbon monoxide. *Ann. N.Y. Acad. Sci.* **174**, 135.

LOVELOCK J. E. (1971) Air pollution and climatic change. *Atmospheric Environment* **5**, 403–411.

LOWE C. R. (1970) Clean air—the balance sheet: the health balance sheet. *Ann. Conf. Natl. Soc. Clean Air*, Southport.

MUNN R. E. and BOLIN B. (1971) Global air pollution—meteorological aspects: a survey. *Atmospheric Environment* **5**, 363–402.

MURRAY M. J. (1970) A preliminary evaluation of atmospheric pollution as a cause of the global temperature fluctuation of the last century. Chapter 2 in Part III of *Global Effects of Environmental Pollution* (edited by SINGER S. F. and REIDEL D.) Dordrecht.

SAWYER J. S. (1971) Possible effects of human activity on the world climate. *Weather* (June 1971).

SCORER R. S. (1957) The cost in Britain of air pollution from different types of source. *J. Inst. Fuel* **11**, 111–111.

U.S. DEPT. OF HEALTH, EDUCATION AND WELFARE PUBLIC HEALTH SERVICE (1964) *Smoking and Health, Report of the Advisory Committee to the Surgeon General of the Public Health Service.*

U.S. DEPT. OF HEALTH, EDUCATION AND WELFARE PUBLIC HEALTH SERVICE (1964) *The Health Consequences of Smoking, A Public Health Service Review.*

U.S. DEPT. OF HEALTH, EDUCATION AND WELFARE PUBLIC HEALTH SERVICE (1968) *The Health Consequences of Smoking, Supplement to the* 1967 *Review.*

153

THE USE OF STANDARDS IN THE ADMINISTRATION OF ENVIRONMENTAL POLLUTION CONTROL PROGRAMS

Morris A. Shiffman, Ph.D., F.A.P.H.A.

THE significance of environmental pollution control standards has changed from their beginnings as simple rules with limited effects on society to their present status as complex functions with multiple effects on major segments of society. The complexities involve matters of scientific fact, but the perplexities are those which involve questions of social and economic costs and of public policy. Standards have been defined as quantitative criteria for social and administrative policy making, whose purpose is to guide or restrain individual or group behavior with an assumed benefit to society as a whole.[1] The particular focus of this study is not to deal with the standard as a technical tool, but rather to explore the use of the standard by the program administrator. The survey is therefore a systematic attempt to record and analyze bureaucratic behavior. A recent report[2] on pollution control emphasizes this approach by noting that, "The issue with respect to standards lies not so much in the concept or use of standards as such, but in how and to what purpose they are used."

The questions that relate to administrative behavior in the use of standards, include:

1. The utility and dis-utility of standards in program planning, program operations, and regulatory activities.
2. The role of standards in decision-making.
3. The administrator's perception of the development of standards.
4. Opinions on the validity of standards.
5. The influence of these opinions on the acceptance and utilization of standards by the administrator.

6. The influence of professional societies on the development and administration of standards.

7. The measure of the authority represented by the standard and the extent of the administrator's dependence on this authority.

8. Intergovernmental problems in the development and administration of standards, including the degree of acceptance of standards promulgated at superior level of government and conflicting requirements of different governmental standards.

9. Variations in the acceptance of standards depending on the source of the standard within government or in the various professional societies.

10. Variations in the utilization of standards relative to the degree of authoritarianism shown by the administrator.

These issues are easily recognizable to environmental health practitioners. Attitudes and opinions on the use of standards have been expressed widely.[3-6] There is an extensive literature which deals with concepts of administrative use and misuse of standards, validity, administrative discretion and responsibility, decision-making, and other significant factors in administrative behavior.[7-14] The concepts of standards as described in this literature were compared with the concepts and attitudes of the practitioners as expressed through the questionnaire study of administrators of environmental control programs.

Methodology

The study was designed to measure the influence of these concepts and attitudes of the administrator on the ways he will use environmental health standards. Attitudes are difficult to define and to measure, but are a vital element in receptiveness to change and predisposition to action. An attempt to make the administrator's attitudes discernible as well as to identify the source of these attitudes is a necessity for the understanding of decision-making. The exploration of this dimension of management was approached through a nationwide questionnaire survey of the relatively small group of environmental health personnel who are responsible for the administration and direction of

environmental health programs, or for the direction and supervision of specialized components of an environmental health program. These persons are assumed to be those who influence the conduct and performance of environmental control programs in public health agencies. Their opinions, attitudes, and perceptions of standards determine the manner in which standards shall be applied in program administration.

The criteria for inclusion of persons in the sampling plan are listed below. All the criteria relating to the type of agency, occupation, and position level had to be satisfied for inclusion in the sampling plan.

A. Persons responsible for the direction of an environmental health program, or for a major component of a program such as sanitary engineering, environmental and food sanitation, industrial hygiene, radiological health, air pollution control. These are programs in which standards have a major role in policy development, administration, and technology.

B. Employment in a state health department or a local health department serving a minimum population of 100,000 and located in an urban area and offering services in several or all the programs specified above. The required degree of comprehensive program coverage was usually reached in those agencies which served urban populations of 200,000 or more. These larger agencies make up the major portion of the sample population for local health agencies.

The federal government has taken an active interest and leadership in the promotion of standards for environmental control, especially in the areas of water pollution, air pollution, food protection, and radiological health. The sample from federal agencies is limited to persons active in these areas.

There are a few cases in which air pollution control is in a division of state or local government other than the health department, and an industrial hygiene program may be found in a state department of labor and industry. The agencies were then included in the sample in these cases. The questionnaire was sent to 447 persons. There were 271 respondents, or 61 per cent of the total mailing, who fully completed the questionnaire. The breakdown in response by the level of governmental agency was as follows:

Agencies	Questionnaires mailed	Replies
State	171	116 or 67.8%
Federal	21	17 or 81.0%
Local	255	138 or 54.1%
	447	271

The 271 respondents who completed the questionnaire gave their primary program area specialty as environmental health administration (110), sanitary engineering (23), environmental and food sanitation (51), industrial hygiene (25), radiological health (29), air pollution control (26), and other (7).

It is necessary to identify those who did not reply to the questionnaire and to learn the reason for this lack of response, insofar as this is possible. There is the requirement for establishing the nature of the nonrespondents and determine if this was due to some basic difference between respondents and nonrespondents. This step was attempted through the examination of (1) letters received from persons who were mailed the questionnaire explaining why they did not return the questionnaire; (2) explanatory notes on questionnaires which were returned without being completed; and (3) checking the completed questionnaires for comments and multiple signatures or other evidence that the questionnaire had been completed by several persons (as an agency consensus) rather than by an individual.

The sampling plan required that separate questionnaires be sent to individuals responsible for separate program components in the same department as well as to the director of the department. In several cases, particularly in the state health departments, the director of the environmental health program responded in the name of the whole agency and for the separate staff members, although separate questionnaires had been sent to these staff members. For example, the following statement is from a completed questionnaire received from the director of environmental health of a state health department:

"These entries are a composite of the views of the staff of the Division. A multiple number of questionnaires were received addressed to various members of the staff, but only one is being returned."

In summary, of the total mailing of 447 questionnaires, complete questionnaires were received from 271 respondents and other replies from an additional 31 persons. In all, replies were received from 302 persons or 68 per cent of those who received the questionnaire. The following analysis of results represent the data from those respondents who returned complete questionnaires. The data was further analyzed for differences in response by generalists and specialists, for program specialty, and for local, state, and federal agencies.

The questionnaire was divided into three sections:

Section A—"General Considerations on the Use and Effectiveness of Environmental Health Standards" contains the questions which deal with the respondents' opinions and perceptions of standards as an integral concept without reference to a specific standard. These results are described in this report on the abstract general concept of the standard and the standards process.

There is, of course, a continuous feedback between knowledge and experience with a specific environmental health standard or standards, and the abstract view of the standards. This essential linkage is recognized in Sections B and C of the survey.

Section B — "Environmental Health Standards in Current Use" deals with specific environmental health standards which have wide application in public health practice.

Section C—"Proposed Environmental Health Standards" which have determined a great deal of interest and some conflict. These include the proposed standards on Ambient Air Quality Criteria, Radiological Health Standards for the General Population, and Microbiological Standards for Foods.

The organization of the survey into these three sections makes possible comparisons based on the linkages between the abstract concept of the standards process, the acquaintance with a specific standard, and opinions on future standards. The results of these linkages and of the data in Sections B and C are only briefly discussed in this report.

Results

A primary purpose of environmental health standards is to provide a tool and yardstick for the program administrator. It is important to assess the administrator's concept of the usefulness of the standard in this respect (Table 1). Program administrators and specialists responsible for environmental pollution control activities in governmental agencies attach a high degree of utility to the use of environmental health standards in program management. However, there is a higher value attached to the use of standards in the administrative processes of enforcement, achieving uniform administration, and measurement of technical problems than in those longer range administrative processes which include program evaluation, program planning, and the maintenance of community and legislative relations. Therefore, the respondent's concept of the utility of standards is as an aid in the day-to-day operational phases of the program rather than with the planning and supportive aspects of the program. There are differences in response between generalists and program specialists in all these administrative processes except for planning. The specialist finds a greater utility for standards in the measurement of environmental health problems, but less use for standards in enforcement and achieving uniform administration.

A dominant theme in the literature of the standards process has been a warning that dependence on standards by the administrator stifles research and development, and fosters a rigidity of attitude to the detriment of the public and industry. The need for flexibility in standards is stressed as well as the necessity for revision with time and technological change. Standards have been cited as impeding the administrator in the use of his best judgment and contributing to a tendency to abdicate responsibility in the performance of his functions. The program administrators and specialists did recognize the possibility of environmental health standards as impediments to effective program operation and administration (Table 2). Almost 50 per cent of the respondents did consider that standards could be deleterious to some extent in these respects; however, very few thought that standards were definitely defective for application in environmental health programs. Over one-third of those replying saw no defect at all in the application of environmental health standards. The results indicate that the program administrators and specialists are sensitive to the possible administrative shortcomings of environmental health standards. However, very few think these are disabling and a large number do not perceive any shortcomings. This result differs then from the view in the literature which emphasizes the presumed defects of environmental health standards in program administration.

Another statement that appears in the literature is that standards which are adopted officially and become an integral part of a legal regulation or code become entrenched and difficult to revise as needs change. The thought is that advisory standards are preferable because of the facility in revising them and because advisory standards do not become rigidly established. There is a marked preference among administrators and specialists for the officially adopted standard (Table 3).

The deleterious effects of these standards on the user have been expressed. Standards are supposed to diminish the judgment and responsibility of the administrator and cause him to evade responsibility through an overdependence on a rigid numerical criterion. Fifty-seven per cent of those replying would themselves reserve the prerogative to

Table 1—Concepts of the usefulness of environmental health standards in program administration

Administrative process	Very useful		Useful		Little use		No response		Total	
	No.	%	No.	%	No.	%	No.	%	No.	%
Program planning	96	35.4	140	51.7	28	10.3	7	2.6	271	100
Measuring environmental health problems	140	51.7	113	41.7	10	3.7	8	3.0	271	100
Enforcement	149	55.0	93	34.3	23	8.5	6	2.2	271	100
Achieving uniform administration	139	51.3	115	42.4	12	4.4	5	1.8	271	100
Public relations	83	30.6	133	49.1	49	18.1	6	2.2	271	100
Program evaluation	93	34.3	127	46.9	43	15.9	8	3.0	271	100

Table 2—Concepts of the defects of environmental health standards in program administration

Administrative objection	Definitely		Some		None		No response		Total	
	No.	%	No.	%	No.	%	No.	%	No.	%
Loss of flexibility in program operation	16	5.9	131	48.3	120	44.3	4	1.5	271	100
Restricting the range of decision-making	26	9.6	136	50.2	103	38.0	6	2.2	271	100
Rigid standards leading to unreasonable requirements	39	14.4	132	48.7	91	33.6	9	3.3	271	100
Established standards stifling new approaches to problem solution	37	13.7	115	42.4	111	41.0	8	3.0	271	100

grant tolerances to the standard based on their judgment of the situation. This represents an expression of administrative discretion. Thirty per cent of the respondents, however, were ready to accept some limitation to the free exercise of their administrative discretion. This result may indicate a conflict between the administrative role which emphasizes regularization of procedures as against the professional value which prizes the use of individual professional judgment. This conflict is related to the real life situation in which the administrator is caught between a public and legislative demand for the rigid imple-

mentation of standards and his professional role which requires a judicious discretion in the application of the standard to the particular situation.

Need for Administrative Interpretation

The same element of conflict is seen in the answers to questions relating to the need for administrative interpretation of the standard (Table 4). Over 60 per cent of the respondents disagreed with a statement that standards should be promulgated in such an exact and definite form that there is little need for administrative interpretation. This

Table 3—Preference for the use of official or advisory standards

Preference scale	No.		%
Advisory standards most preferable		32	11.8
	78		28.8
Advisory standards preferable		46	17.0
Equal preference		69	25.5
Official standards preferable		65	24.0
	120		44.3
Official standards most preferable		55	20.3
No response		4	1.5
		271	100.0

would represent the professional value and is the point of view supported by much of the literature on the standards process. The surprising feature in these responses, however, was the large group of respondents (35%) who agreed that there was little need for administrative interpretation in the implementation of standards. This latter view represents a particularly rigid attitude in the application of standards.

The validity of the standard is an important concept in the standards process. The acceptance and use of the standard may be influenced by the user's perception of its validity. In this study validity was defined in terms of the objective procedures for developing and promulgating the standard, including the availability of published information and criteria which are the basis for the standard. The primary factor in the user's acceptance of the standard was his confidence in the organization which was the source of the development of the standard (Table 5). The second factor chosen was the respondent's knowledge of the professional reputation of the committee members who were charged with the development of the standards. There was some difference between gen-

eralists and specialists in this respect. The generalists had a higher confidence in the organization source (68% vs. 54%), while the specialists emphasized the professional reputation of the committee members (23% vs. 14%). Comparison of these results with the literature is also interesting. The standards user has a very high confidence in the expert committee which develops the standards (Table 6). On the other hand, members of the expert committees tend to be somewhat apologetic in their description of the standards process, noting the compromises, inadequacies, and judgmental elements involved in the setting of a standard.

The development of standards requires a complex interrelationship and cooperative effort between members of governmental agencies, professional societies, and industry. Despite this coordinated effort, each standard bears the imprint or label of some government organization, professional society, or other identifiable source. This primary source label identifies the standard in the mind of the user and influences his perception of the utility and validity of the standard. There also may be differing perceptions attached to a standard which derives from a federal, state, or local source. One of the areas of conflict in the development of environmental pollution control standards has been, in different opinions, as to the most appropriate role for federal, state, or local promulgation for air quality standards and water quality standards. An important point in the explanation of the standards process is to delineate the role and relative influence of the source which is the prime initiator of the standard. The initiation and development of a standard rest most significantly within one group, be it a governmental agency or a professional society. The next series of questions was designed to test the degree to which differences in the source of the standard influenced the program administrator.

It may be assumed that almost all respondents are members of one or several professional societies. This assump-

Table 4—Concepts of administrative discretion and responsibility in the use of environmental health standards

a. Standards should be promulgated in such an exact and definite form that there is little need for administrative interpretation

Agreement scale	No.	%
Definitely agree	27	10.0
Tend to agree	66	24.4
Don't know	4	1.5
Tend to disagree	84	31.0
Definitely disagree	88	32.5
No response	2	.7
	271	100.0

b. Administrators utilize environmental health standards to minimize their responsibility for making decisions

Definitely agree	17	6.3
Tend to agree	63	23.2
Don't know	24	8.9
Tend to disagree	94	34.7
Definitely disagree	71	26.2
No response	2	.7
	271	100.0

tion is supported by the fact that the names of the respondents were derived in part from professional directories or membership lists. Furthermore, many of the respondents have attained a degree of professional distinction and it is usual for persons at this level of attainment to be active in their professional societies. Therefore, it would expected that their values and belief system would adhere most closely to those of the professional societies. However, standards developed by governmental agencies rather than by professional societies were preferred (Table 7). The answers offer some insight into the integration of technical and professional personnel into a bureaucratic organization. The scientist or technologist who is a member of a bureaucracy is subject to a dual social control from his professional reference group and from the employing agency.

There is interest among sociologists[15,16] as to whether professionals derive their social control and express their loyalty to their professional commitments above their organizational commitments. These responses demonstrated that the specialist was more likely to prefer the standards of professional societies than was the generalist. However, both groups showed a definite choice for the standard of the governmental agency over the professional society standard.

This line of inquiry was pursued from several points of view. In each case, the overriding opinion was that the standard derived from a government agency source was the more effective, the more useful in an operational sense. and best suited to the needs of the respondent than the standards derived (Tables 8, 9). These results may indicate that the user's perceptions of standards from a professional society source are as devices which serve his administrative needs and his values as a member of the bureaucracy rather than the professional value.

The relative influence on program operations of standards issued by federal, state, and local governments was explored in the next series of questions. In appraising the relative merit of these governmental sources, the respondents from local agencies have a generally favorable opinion of state standards as measured by the extent of need for modifying these standards for local use. The state standard was considered more applicable than the federal standard or the professional standard for application at the local level (Table 10).

Stringency of the Standard

Another important variable is the user's perception of the stringency of the standard. Approximately two-thirds of the respondents felt that federal or state health department standards were stringent enough for local use (Table 11). A small number (23.6% for state standards and 14.4% for federal) felt that these standards were not stringent

Table 5—Concepts of the bases for acceptance of a standard as authoritative

Bases for acceptance	First choice No.	%	Second choice No.	%
Confidence in the organization which is the primary source for the development of the standard	166 (100)* (66)†	61.3	52 (26) (26)	19.2
Knowledge of the professional reputation of individuals on the committee which developed the standard	49 (21) (28)	18.1	92 (47) (45)	33.9
Personal acquaintance with individuals on the committee	6 (3) (3)	2.2	11 (6) (5)	4.1
Recognition of other organizations represented on the committee	9 (2) (7)	3.3	74 (47) (27)	27.3
Other factors	29 (15) (14)	10.7	13 (7) (6)	4.8
No response	12 (7) (5)	4.4	29 (15) (14)	10.7
Total	271	100.0	271	100.0

* Generalist.
† Specialist.

Table 6—Perceptions of the scientific validity of environmental health standards developed by committees of recognized experts

Opinion on validity	No.	%
Mostly valid	132	48.7
Frequently valid	116	42.8
Occasionally valid	16	5.9
Rarely valid	0	0.0
Don't know	4	1.5
No response	3	1.1
	271	100.0

Table 7—Comparison of the influence of standards developed by governmental agencies and professional societies on program operations

Extent of influence	Standards developed by			
	Government agencies No.	%	Professional societies No.	%
Very much	145	53.5	52	19.2
Much	89	32.8	101	37.3
Some	24	8.9	90	33.2
Little	3	1.1	14	5.2
None	0	0.0	2	0.7
No response	10	3.7	12	4.4
Total	271	100.0	271	100.0

enough. There is a significant difference in the responses of personnel in state and local health agencies. Three-quarters of respondents from state agencies thought that both state and federal standards were acceptable in this respect for local use. The local health group agreed insofar as the acceptability of federal standards; however, one-third of the local personnel thought that state health standards were lacking in stringency. A reason for this reply may be in the different level of familiarity of local personnel with the federal and state standards. The local personnel may be more familiar with the use of state standards and are more knowledgeable as to their operational characteristics in practice.

One of the cautions in the literature is that administrators use numerical criteria in a mechanistic manner, especially in enforcement proceedings. The respondents were asked at what point they would initiate legal enforcement following the first violation, second violation, repeated violations, or no enforcement at all based on violation of the standard. Only 6 per cent of the respondents would initiate enforcement following the first violation. Half of the respondents would only initiate enforcement following repeated violations. These data show that there is no tendency by the practitioner to use standards in a mechanical way and that enforcement should only follow repeated violations. It is interesting to note that the specialists were more likely to reject numerical criteria as a cause for legal enforcement. Ten per cent of the specialists indicated that there should be no enforcement action based on the violation of a standard, while only 3 per cent of the generalists would take extreme positions.

As noted, another section of the questionnaire dealt with a group of specific environmental health standards in current use. These findings are not explored in detail here; however, some points of interest are summarized. The utility of these standards in practice was given a high rating, especially for the application of these current standards in enforcement and in the achievement of uniform patterns of administration. In common with the administrators' general concept of the use of standards, planning is given a lower degree of acceptance. All three specific standards were given high ratings insofar as the users' opinion on their practicability in program operations.

Over 90 per cent of the respondents indicated that the Public Health Service Drinking Water Standards and the bacteriological standards of the Recommended Milk Code were practical to administer. Eighty-six per cent of those familiar with Industrial Hygiene Threshold Limits felt these were practical to administer.

Future Use of Standards

A section of the study dealt with the future use of standards to meet new problems in environmental health. This section is intended to translate the respondents' general concepts and opinions on current standards into preferences for specific attributes such as the source, format, and desirability for the development of ambient air quality control criteria, radiological health standards, and microbiological standards for foods. The respondents expressed a high degree of need for such standards. This reinforces the conclusion that environmental health personnel have a strong dependence on and attach a high value to the utility of standards in program operation. The respondents also attach a higher acceptance of officially adopted standards over standards adopted in the advisory form. It is interesting to note that though there was a strong expressed need for these proposed standards, there were reservations as to the possibility of developing valid standards for the control of these pollution problems. This lesser confidence in the validity of these standards as compared to current standards may stem from the conflict surrounding the development of some of these particular standards. The lesser confidence in the validity of these standards did not affect the practitioners' expression of need for

Table 8—Comparison of the relative effectiveness of standards issued by governmental agencies and professional societies in program administration

Source of the standard	No.	%
Government most effective	43	15.9
Government more effective	132	48.7
Equal effectiveness	74	27.3
Professional society more effective	7	2.6
Professional society most effective	2	.7
No response	13	4.8
Total	271	100.0

Table 9—Comparison of the influence of government and professional societies standards as guides in the development of community standards

Influence of the source of the standard	No.	%
Government most important	56	20.7
Government more important	81	29.9
Equal importance	93	34.3
Professional societies more important	11	4.1
Professional societies most important	8	3.0
No response	22	8.1
Total	271	100.0

Table 10—Opinions on the need for the modification of standards to suit the requirements for local health department use

Need for modification	Source of the standard					
	Federal agencies		State agencies		Professional societies	
	No.	%	No.	%	No.	%
Extensive	10	3.7	8	3.0	24	8.9
Some	178	65.7	125	46.1	171	63.1
Little	54	19.9	99	36.5	46	17.0
No change	12	4.4	21	7.7	7	2.6
No response	17	6.3	18	6.6	23	8.5
Total	271	100.0	271	100.0	271	100.0

Table 11—Opinions on the strictness of
public health service and state health
department standards in reference to
their application in the local community

Stringency	Source of the standard			
	Public Health Service		State health department	
	No.	%	No.	%
Not stringent enough	39	14.4	64	23.6
Acceptable	194	71.6	176	64.9
Too stringent	13	4.8	2	.7
No response	25	9.2	29	10.7
Total	271	100.0	271	100.0

such a standard for program operations.
It may be that the concept of validity
is more important to the standards-setter
and the expert than it is to the administrator and program specialist. A decreased perception of validity does not
seem to impede the proposed operational
use of the standard by the program administrator and specialist.

There was an overwhelming demand
for a governmental agency source for
these proposed standards rather than a
professional society—with the federal
government as the most preferred source
followed by the state, and with the local
government as the least preferred governmental source.

A standard is not an isolated pro-
nouncement. Social forces are exerted
to initiate the development of standards.
The development process involves the in-
terplay of technical, economic, legal, social, and political factors. By a similar
token, standards once developed do not
exist in the abstract. They are applied
in a real world by human beings. The
focus on attitudes and behavior in the
implementation of standards in environmental pollution control programs is a
useful approach to the exploration of the
origins of administrative behavior.

REFERENCES

1. Wolman, Abel. Concepts of Policy in the Formulation of So-Called Standards of Health and Safety. J. Am. Water Works A. 52:1343, 1960.
2. National Academy of Sciences-National Research Council. Waste Management and Control. Publ. No. 1400, Washington, D. C., 1966.
3. U. S. Congress, Joint Committee on Atomic Energy. Hearings, Radiation Pro-tection Criteria and Standards: Their Basis and Use, 86th Congress, 2nd Ses-sion, 1960.
4. Thomas, Harold A., Jr. The Animal Farm: A Mathematical Model for the Discussion of Social Standards for the Control of the Environment. Quart. J. Economics 77:144, 1963.
5. Kneese, Allen V. Comments on Harold Thomas' Animal Farm. Prepared for pre-sentation before the joint meeting of the Subcommittee on Water Supply and Waste Disposal of the Committee on Sanitary Engineering and Environment, NAS-NRC, Washington (May 7), 1963 (mimeo.).
6. Goldsmith, J. R. Bases and Criteria for Air Quality Standards. J. Air Pollution Control A. 14:22, 1964.
7. Irish, D. D. Evolution of Our Concept of Standards. Arch. Environmental Health 10:546, 1955.
8. Shuval, Hillel I. The Uses and Misuses of Environmental Health Standards. A.J.P.H. 54:1319, 1964.
9. Levine, M. Facts and Fancies of Bacterial Indices in Standards for Water and Foods. Food Technol. 15:29, 1961.
10. McKee, J. E., and Wolf, H. W. (eds.). Water Quality Criteria (2nd ed.). Sacra-mento: California State Water Quality Control Board, 1963.
11. MacKenzie, V. G. Air Pollution Standards. Arch. Environmental Health 2:224, 1961.
12. McCord, C. P. Man and His Environment —A Conference Resume. Indust. Med. & Surg. 30:354, 1961.
13. McGaughey, P. H. Folklore in Water Quality Standards. Civil Engineering (June), 1965.
14. Davis, Philip J. The Criterion Makers: Mathematics and Social Policy. Am. Scientist 50:254A, 1962.
15. Gouldner, Alvin W. Cosmopolitans and Locals. Administrative Science Quart. 2: 281, 1957.
16. Blau, Peter M., and Scott, Richard W. Formal Organization, A Comparative Ap-proach. New York: Chandler Publishing, 1962.

164

Air Pollution Control Training in Colleges and Universities in the United States

S-11 Education and Training Committee Survey Report No. 2

Principal Author: Harold M. Cota

This paper was prepared to provide an overview of current air pollution training in the United States for the S-11 Education and Training Committee, APCA. An earlier report by Sholtes[1] reviewed the wide range of training going on in 1966. The number of programs has significantly increased; therefore, the present study has been focused on training efforts at colleges and universities leading to academic degrees.

Statements indicating a need for manpower trained in air pollution control are frequently cited. To meet this challenge, many schools now offer training in air pollution control. Federal programs partially support some of these efforts. The objective of this study was to determine the extent of the training now going on in the United States. Of particular concern was the (a) type of program, (b) background experience of students in the program, and (c) eventual placement of the students.

Questionnaires were sent to all schools listed in the Journal of Engineering Education Directory or with faculty members in APCA or listed in data obtained from the Air Pollution Control Office-EPA. Response from 96 schools that had programs identified with air pollution control training was received.

Academic Programs

Complete lists of all schools with instructional effort in air pollution control are given in Appendix I and II, respectively. To determine the scope and particular direction of each program requires a detailed study of each specific curriculum. For the purposes of this study, certain key factors were used instead to characterize the work offered. Most programs were centered in departments of Civil, Chemical, Environmental, or Mechanical Engineering. Sixty-three percent of 466 faculty involved in air pollution training in 72 schools responding are identified with these four departments. The remaining participating faculty are distributed as follows: Biological Sciences—11%, Meteorology-8.7%, Chemistry—6.6%, Public Health—5.7%, Public Administration—1.8%, and others– 3.2%.

The courses taken by students in these programs covered a wide range of subject matter. The types of courses offered were reported by 67 schools and are shown in Table I.

In 63 out of 72 programs, non-degree candidates may enroll in air pollution courses. The five community colleges

Table I. Courses offered in training programs in 67 schools.

Course	Number of schools
Introduction to Air Pollution	56
Sampling and Analysis	44
Meteorology	46
Control Technology	40
Legal Aspects	24
Community Planning	21
Administrative Procedures	12
Public Information—Community Relations	11
Statistics	44
Other	23

who completed questionnaires are not included in the above statistics.

Financial assistance was reportedly available at 33 schools from several sources. This information is summarized in Table II.

Students

The academic background of students entering the air pollution programs varied. An overall distribution of background is presented in Figure 1. Of these students, many had some experience relevant to air pollution control. Up to 25% of the students had experience with NAPCA, (now APCO-EPA), up to 40% with control agencies, and up to 50% with industry, depending on the school. Apparently over 70% of those in the programs that furnished data are graduate students.

Table II. Number of students with various types of financial assistance.

Source of assistance	Number of students and degree		
	BS	MS	PhD
Part-time A.P.C. agency	1	5	1
Summer A.P.C. agency	11	7	3
Summer work with industry in A.P.C. areas	18	9	4
Fellowships (Industry)	6	10	13
Assistantships	4	50	27
Fellowships (Foundation)	1	9	9
PHS Traineeships		113	21
Total	41	203	78

Table III. Geographical distribution of air pollution control jobs accepted.

Location	Degree Obtained		
	BS	MS	PhD
On West Coast	26	28	5
On East Coast	2	84	17
In Central States	1	98	11

Graduates

Although a total of 2456 students were reported taking course work in air pollution control, the number getting in-depth training and later working as

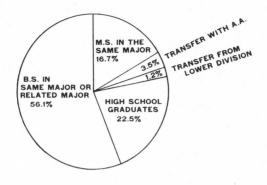

57 SCHOOLS REPORTING

Figure 1. Distribution of academic background of students entering air pollution programs.

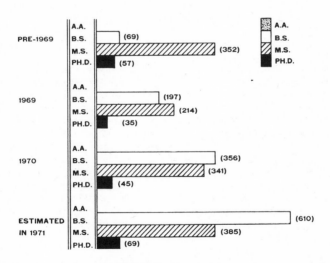

57 SCHOOLS REPORTING

Figure 2. Number of students graduated with training in air pollution control.

167

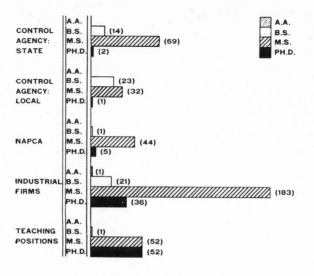

CONTROL AGENCY: STATE
- A.A.
- B.S. (14)
- M.S. (69)
- PH.D. (2)

CONTROL AGENCY: LOCAL
- A.A.
- B.S. (23)
- M.S. (32)
- PH.D. (1)

NAPCA
- A.A.
- B.S. (1)
- M.S. (44)
- PH.D. (5)

INDUSTRIAL FIRMS
- A.A. (1)
- B.S. (21)
- M.S. (183)
- PH.D. (36)

TEACHING POSITIONS
- A.A.
- B.S. (1)
- M.S. (52)
- PH.D. (52)

Legend:
- A.A.
- B.S.
- M.S.
- PH.D.

41 SCHOOLS REPORTING

Figure 3. Number of students completing an air pollution program and placed in related work.

professionals or semi-professionals is much less. The number of students graduating in recent years is an indicator of the latter. Figure 2 shows how the number of graduates at all levels has grown since 1969 at 57 schools.

Figure 2 does not include 308 graduates classified as receiving a degree other than the B.S., M.S., or Ph.D.

Of considerable interest is where the various graduates are now working. Figure 3 correlates the total number of graduates taking initial employment in the air pollution field at 41 schools. Figure 3 does not include 113 who were placed but did not have the degrees mentioned above.

Information on the geographical location of air pollution control jobs was limited. The data from 28 schools reporting are included in Table III.

Conclusions

1. This study reveals that more than 96 schools in the United States are currently involved in air pollution control training. Faculty involvement was estimated at 466 by 73 schools reporting.

2. It was estimated that 610 BS, 385 MS, and 69 PhD students would graduate in 1971 with some training in air pollution control from 57 schools. The interpretation of training was taken broadly.

3. At this time, there are more students with Masters Degrees taking initial employment in the air pollution control field.

4. The level of support is small and needs to be encouraged at all levels from many sources.

The data on which the above conclusions were based are not complete. All schools with programs were not able to complete the questionnaire. Often the information desired was not available. Based on the experience gained in this report, the Education and Training Committee, S-11, is encouraged to update these estimates on a regular basis. In addition, community college and specialized training programs should be followed. The author invites comments on this report.

Studies have been made to estimate manpower needs in the governmental and private sectors.[2] One conclusion from these estimates as well as observation is that there is a growing need for manpower committed to working toward the solution of air pollution problems. This has been the basis for a significant commitment of faculty, students, and resources. Determining what the manpower needs actually are is another area where some effort must be placed.

References

1. Sholtes, R. S., "Report of Education and Training Committee, S-11," *J. Air Poll. Control Assoc.* **16** (11), 610 (1966).
2. "Manpower and Training Needs for Air Pollution Control," Report to the President and Congress by the Secretary of Health, Education, and Welfare, June 1970.

Appendix I. Air pollution control training at colleges and universities in the United States.

State	College or University	Program Concerned In	# Courses	Degrees
Alabama	Samford U.	None	5	None
	U. of Alabama	ChE	4	BS, MS
Arizona	U. of Arizona	CE	5	BS, MS, PhD
Arkansas	U. of Arkansas		2	MS
California	Calif. Inst. of Tech.	EnvE, ChE	18	Post-Doc, PhD
	Cal. State—Long Beach		1	
	Cal. St. Poly-San Luis Obispo	EnvE	6	BS, MEngr.
	Sacramento State	CE, ME		BS, MS
	San Jose State	Grad	8	MS
	Stanford U.	ME		BS, MS, PhD
	U. of Calif.—Berkeley	ME	2	MS, PhD
	U. of Calif.—Davis	CE	5	MS, PhD
	U. of Calif—Irvine	EnvE	5	
	U. of Calif.—Los Angeles	Engr	10	MS, PhD
	U. of Calif.—Riverside[1]			
	U. of Southern Calif.	APC Inst	8	M.P.A.
Colorado	Adams State College[2]			
	Colorado State U.		1	
	U. of Colorado	Chem		
	U. of Denver	Engr		
Connecticut	Yale U.	PH	8	M.P.H.
Delaware	U. of Delaware			
Florida	U. of Florida	Engr	10	MS, PhD
Georgia	Columbus College			BS
	Georgia Inst. of Tech.	ChE	6	
	U. of Georgia	AgE	1	
Hawaii	U. of Hawaii	PH	6	MS, PhD
Illinois	Bradley U.		1	
	Northwestern U.	Engr	3	MS, PhD
	Southern Illinois U.	Engr		BS, MS
	U. of Illinois—Chicago Circle	Energy E	18	BS
	U. of Illinois—Urbana	Engr		BS, MS
Indiana	Purdue U.	CE	10	MS, PhD
	Rose Polytechnic Inst.	Bio, CE ChE	2	BS
	U. of Notre Dame	CE	9	BS, MS, PhD

Appendix I. *(cont.)*

Iowa	Iowa State U.	CE	2	BS, MS
	U. of Iowa	EnvE	1	
Kansas	Kansas State U.	ME, ChE	3	MS, PhD
	U. of Kansas	Env H	1	MS, PhD
Kentucky	U. of Kentucky	Engr, ChE	7	BS, MS, PhD
	Western Kentucky U.	E Tech	3	BS
Louisiana	Lousiana State U.	Chem		
	Louisiana Tech. U.	CE	1	
	LSU—Baton Rouge	EnvE	4	
	Tulane U.			
Maine	U. of Maine	Engr	1	MS
Maryland	Johns Hopkins U.	EnvE	5	MS, PhD
	U. of Maryland	CE, ChE, Met	7	BS, MS, PhD
Massachusetts	Harvard U.	PH	10	BS, MS, PhD
	Mass. Inst. of Tech.	Engr		MS, PhD
	Northeastern U.	CE	6	BS, MS
	U. of Massachusetts	ME, CE, ChE	4	
		PH		
Michigan	Ferris State College	EnvH	7	BS
	U. of Detroit	Engr	6	
	U. of Michigan	Engr, PH	5	MS, PhD
Minnesota	Bemidji College			
	U. of Minnesota	PH	4	MS, PhD
Mississippi	Mississippi State U.	CE	1	
Missouri	St. Louis U.[2]			
	U. of Missouri—Rolla	CE	2	BS, MS, PhD
	Washington U.	Engr Sci	7	MS, PhD
New Hampshire	U. of New Hampshire	ChE	2	BS, MS, PhD
New Jersey	Newark College of Engr.	CE, EnvE	3	MS
	Rutgers CAES	Bio		
New Mexico	N. M. Inst. Mining & Tech.	EnvE		
New York	Cooper Union	Grad	5	MS, PhD
	Cornell U.	Engr	3	
	New York U.	Engr, Sci	5	
	Rensselaer Poly. Inst.—Troy	EnvE	5	MS, PhD
	SUNY—Potsdam	CE, ChE	4	BS, MS
	Union College	ME	4	BS
North Carolina	North Carolina State U.	ChE	4	BS, MS
	U. of N. C.—Chapel Hill	PH, EnvE	10	MS, PhD
Ohio	Bowling Green State U.[4]	IT		
	U. of Cincinnati	Engr	9	MS, PhD
Oklahoma	Oklahoma State U.	CE	1	
	Tulsa U.	ME, ChE	1	
	U. of Oklahoma	CE, EnvH	7	MS, PhD
Oregon	Oregon State U.	ME	9	BS, MS, PhD
	Oregon Tech. Inst.	EnvH	4	AA, BS
	Portland State U.	Engr, Sci	6	BS, MS
Pennsylvania	Carnegie Mellon U.			
	Drexel U.	EnvE-Sci	8	MS, PhD
	U. of Pittsburgh	PH	7	MS, PhD
	Pennsylvania State U.	Air Env	10	AA, BS, MS, PhD
Utah	U. of Utah	Bio	6	MS, PhD
Virginia	U. of Virginia	Engr, Sci		MS
	Virginia Poly. Inst. & St. U	CE		BS, MS, PhD
Washington	U. of Washington—Seattle	CE		
	Washington State U.	CE, Env Sci	9	
West Virginia	West Virginia U.	CE	5	MS, PhD
Wisconsin	Marquette U.	CE, Med	2	
	U. of Wisconsin—Madison	Met, ME	1	
		ChE, CE		
Wyoming	U. of Wyoming	EnvH	3	

[1] The Statewide Air Pollution Center provides instructors for several air pollution related courses in other departments and frequently has graduate students conduct their research at the center.
[2] Currently in final stages of having Environmental Science curriculum approved.
[3] Currently working on programs that involve air pollution meteorology.
[4] In planning stage.

Appendix II. Air pollution control training at community colleges
in the United States.

State	College or University	Program Concerned In	# Courses	Degree
California	El Camino College	Nat Sci	2	AA
Florida	Santa Fe Junior College	Engr	5	AA
Maryland	Charles County Com. College		4	AA
Michigan	Genesee Community Col.—Flint	EnvCont Tech, AP	3	AA
New York	Broome Tech. Com. Col.	EnvH	1	AA
	Corning Com. College	Bio, Chem		

Abbreviations Used in Appendix I and II

Air Env	Air Environ. Studies Center
AgE	Agriculture Engineering
APC Inst	Air Pollution Control Institute
Bio	Biology
CE	Civil Engineering
ChE	Chemical Engineering
Chem	Chemistry
Engr	Engineering
EnvE	Environmental Engineering
EnvH	Environmental Health
EnvR	Environ. Resource Engr.
EnvSci	Environmental Science
E. Tech	Engineering Technology
Grad	Graduate School
IT	Industrial Technology
ME	Mechanical Engineering
MEngr	Master of Engineering
Med	Medicine
M.P.A.	Master of Public Admin.
M.P.H.	Master of Public Health
Met	Meteorology
NatSci	Natural Science
PH	Public Health

171

Several schools originally included in the paper 71-166, presented at the 64th Annual Meeting of the Air Pollution Control Association, at Atlantic City, New Jersey, June 1971 did not appear in the JAPCA article. These schools are listed below. They each have air pollution control training activities.

STATE	COLLEGE OR UNIVERSITY	PROGRAM CONCENTRATED IN	# COURSES	DEGREES
*NEW YORK	City University of N.Y.	EnvE	5	M Engr, MS
SOUTH CAROLINA	Clemson U.		2	MS,PhD
SOUTH DAKOTA	S.D. Sch. of Mines & Tech			
TENNESSEE	U. of Tenn., Knoxville	CE, ChE	3	BS, MS
	Vanderbilt U.		7	MS, PhD
TEXAS	Lamar St. Col of Tech	Env Sci	5	BS
	Rice U.		2	
	Southern Methodist U.		4	
	Texas A & M U. (11)	CE, Met	4	MS, PhD
	U. of Houston	ChE	3	BS,MS

* Additions received after publication of article

Petroleum fuels and cleaner air

GS Parkinson

In the UK, the use of oil products has increased rapidly since 1945 (Table 1) and petroleum fuels already supply about half of the UK's total energy requirement. This vastly increased use of oil products of all kinds is part of a worldwide pattern, for traditional fuels were not available at the price or in the volume necessary to sustain post-war demands. For many processes, petroleum fuels also have technical advantages which have resulted in process improvement and consequent cost saving. In the future, the demand for petroleum fuels will go on increasing.

Physical characteristics of the principal classes of petroleum fuels are given in Table 2. Propane and butane are gaseous at normal temperatures and are held in the liquid state by storage in containers under moderate pressure. Gasoline, kerosine and diesel fuels are distillates, but fuel oils—also known as residual fuels—contain a variable proportion of residue boiling above 370 °C, which is the limit of atmospheric distillation.

Oil products are composed mainly of hydrocarbons and on combustion produce large quantities of carbon dioxide and water vapour. Smaller amounts of sulphur compounds, present in an organic form, give sulphur oxides, and nitrogen oxides are formed by fixation of atmospheric nitrogen. The way in which the fuel is burned has a significant effect on the types of combustion products, *e.g.* the internal combustion engine, working on a cyclic principle, produces a number of products of partial oxidation. When the products of combustion from oil fuels are discharged to the atmosphere, they contribute to the general atmospheric pollution. The main pollutants are considered here.

Sulphur oxides

Sulphur is present to some extent in most primary fuels: in refined petroleum products, amounts can range from negligible to about 4 per cent depending on the class of fuel and the type of crude oil used. Many of the crudes now available from North and West African fields are relatively low in sulphur and

Table 1. UK consumption of petroleum fuels (thousand long tons).

	1938	1950	1969
Propane and butane	2	30	1 189
Aviation fuels	113	458	3 298
Motor gasoline	4830	5 195	13 232
Kerosine	721	1 363	2 272
Automotive diesel	387	1 034	4 791
Industrial gas oil	797	1 595	10 349
Fuel oils	628	3 093	33 393
Total	**7478**	**12 768**	**68 524**

already represent a substantial proportion of UK refinery intake.

On combustion, most of the sulphur is oxidized to SO_2 with about 3 per cent being converted to SO_3.[1] If the amount of air used in the combustion process is kept to a minimum, SO_3 formation can be reduced to very small proportions;[2] but this is only feasible in the larger well automated industrial burners using residual fuels. Nevertheless much can be done with smaller sized units to prevent undesirable side effects such as the formation of acid smuts. Correct chimney design is one field in which oil industry research has pioneered a line of thinking directed to overcoming this problem.

In 1965 about 6.5 Mton SO_2 from the combustion of all types of primary fuel was discharged into the atmosphere. By 1968, this figure had decreased to 6.14 Mton although the total energy consumption in the UK had risen during the period and was then running at about 300 Mton coal equivalent. The amount of SO_2 emitted now seems to have passed its peak—for the forseeable future, at least—despite a continuing increase in total energy demand.[4] The reason for this is that coal has now largely been replaced by oil and, to a lesser extent, by gas. Of the total SO_2 emitted, about a third is discharged at very high level from power station chimneys some of which are 1000 ft high. At this level, the SO_2 is quickly dispersed into the atmosphere and does not contribute significantly to the ground level concentrations measured in the vicinity.

The 1956 Clean Air Act urged local authorities to set up smoke control areas. A major source of smoke, and of SO_2, was the domestic fire burning raw coal and, as a result of the Act, there has been a significant countrywide reduction in smoke, and also in SO_2 at ground level. Much of the latter reduction is due to the replacement of solid fuel by distillate oil, gas or electricity for domestic and industrial heating. A countrywide network of measuring stations operated by the DTI Warren Spring Laboratory has monitored the ground level concentration of SO_2 in the atmosphere for many years: between 1958 and 1968 the annual mean concentration steadily declined from 170 μg m^{-3} to 110 μg m^{-3}.[5]

The positive evidence that both total SO_2 emissions and ground level concentrations have been reduced is most encouraging: the increasing use of petroleum fuels has played, and will continue to play, a useful role. In the next few years, it should be possible to reduce the average sulphur content of residual fuel oils—currently about 2.5-3 per cent by weight—still further by selection of lower sulphur crude oils.

Technically, it is feasible to reduce SO_2 emissions even further by de-sulphurizing residual fuel oils. Two years ago it was estimated that this process would cost 15 shillings per ton of treated fuel for each 1 per cent reduction in sulphur[4] but by now this must be in excess of £1 per ton.

Alternatives to oil desulphurization have been investigated in which sulphur oxides are removed either during the combustion process or from the gases immediately afterwards and before discharge to the atmosphere.[6,7] A good deal of optimism regarding these processes is often generated because of the variable market price of sulphur which might make them economically

Table 2. Some typical physical properties of oil fuels.

Fuel	SG at 15.5 °C	Kinematic viscosity (centistokes) (°C)	Boiling range	Carbon	Ultimate analysis (% wt) Hydrogen	Sulphur	Trace metals*
Commercial propane	0.51	0.15 (15.5 °C)	−45	82.8	17.2	Negligible	—
Commercial butane	0.58	0.23	−5	83.5	16.5	Negligible	—
Motor gasoline	0.75	0.8	30−180	85.5	14.4	<0.1	—
Domestic kerosine	0.78	2.0	150−260	85.8	14.1	0.1	—
Industrial kerosine	0.79	2.0	150−260	85.8	14.1	0.1	—
Automotive diesel fuel	0.84	5.0	180−350	86.5	13.2	0.3	—
Industrial gas oil	0.83	5.0	180−350	86.2	13.2	0.6	—
Light fuel oil	0.93	12 (82.2 °C)	180−>370	85.7	11.7	2.5	<0.1
Medium fuel oil	0.95	28	180−>370	85.7	11.6	2.6	<0.1
Heavy fuel oil	0.96	60	180−>370	86.0	11.2	2.7	<0.1

* Mainly vanadium, nickel and sodium.

Table 3. Influence of driving conditions on average car exhaust emissions.

Compound	Idling	Acceleration	Cruising	Deceleration
Hydrocarbons, as hexane (ppm)	800	510	185	5000
Hydrocarbons range, as hexane (ppm)	300−1000	300−800	250−550	3000−12 000
Acetylene (ppm)	710	170	178	1096
Aldehydes (ppm)	15	27	34	199
Nitrogen oxides, as nitrogen dioxide (ppm)	10−50	1000−4000	1000−3000	5−50
Carbon monoxide (percentage)	4−9	1−8	1−7	3−4
Carbon dioxide (percentage)	10.2	12.1	12.4	6.0
Oxygen (percentage)	1.8	1.5	1.7	8.1
Exhaust gas volume (standard ft^3 min^{-1})	8	60	35	8
Exhaust gas volume range (standard ft^3 min^{-1})	5−25	40−200	25−60	5−25
Exhaust gas temperature, at catalytic device inlet (°C)	150−300	450−700	400−600	200−400
Rate of consumption (gal h^{-1})	0.495	3.39	1.94	0.495
Unburnt fuel, as hexane (weight per cent of fuel supplied)	2.88	2.12	1.95	18.0

viable. But this is far from being likely. As a starting material, sulphur dioxide is not technically attractive and the process would be carried out in a number of units all over the country—many of which would only produce small amounts of elemental sulphur or sulphuric acid—often in areas remote from any possible market demand. Economic assessments have always shown that this type of desulphurization can never hope to pay for itself and comes nearest to doing so where the quantity of fuel used is very large: in such cases the need is far less pressing because the flue gases are discharged at very high level.

Whatever the process, the cost of sulphur removal would fall on the fuel user and so, ultimately, the community. It could represent a very large additional burden on the country's economy and its justification must be considered not only in the light of our present knowledge regarding SO_2 in the atmosphere but also on rational predictions that can be made from this knowledge where existing data is inadequate. Sulphur oxides from combustion of fossil fuels contribute only an estimated one-third to the total global atmospheric burden although, of course, this proportion will be relatively greater over industrialized areas. The rest of the airborne sulphur, that is, apart from that derived from combustion processes, is present as SO_2 from biological decay (45 per cent) and sulphate from sea spray (25 per cent).[4]

The present or predicted SO_2 levels in the UK atmosphere do not raise any undue medical concern amongst those best qualified to judge. There is a health hazard from sulphur oxides in concentrations considerably higher than the average national survey figures, and in combination with particulate matter much of which comes from domestic fires burning raw coal. The best way of removing this health hazard is to speed up the establishment of smoke control areas.

The case for sulphur removal rests more on damage to amenity and in particular to plant life and buildings. Some species, for example lichens, are sensitive to extremely low concentrations of SO_2 but it is questionable whether there is significant damage to most normal forms of vegetation except in isolated areas where SO_2 concentrations are very high. Sulphur oxides certainly cause some damage to the stonework of buildings, but hydrochloric acid from the combustion of solid fuel may also play a role—it is estimated that the amount of HCl discharged to the atmosphere will rise from 73 000 tons in 1950 to about 200 000 tons by 1975.[4]

The oil industry has been active in trying to fill the gaps in our knowledge regarding atmospheric SO_2. BP conducted a large-scale investigation to correlate SO_2 emissions at various heights in a large town, i.e. domestic, low and medium industrial emitters, with the measured ground level concentrations.[9] The Institute of Petroleum commissioned a literature survey to review critically the current state of our knowledge of the eventual fate of SO_2 in the atmosphere, particularly that discharged at high level, and found that one of the main areas still requiring investigation was the role played in SO_2 oxidation by other atmospheric constituents especially aerosols.[10] The Institute, with the CEGB and the DTI, has set up a study group 'to consider what ground level, aerial and laboratory studies can be made to obtain quantitative evidence regarding the movement and fate of sulphur oxides emanating from all kinds of fossil fuel burning in the UK'. Preliminary recommendations should be available soon, but it is already clear that new analytical techniques will be needed and some method will have to be found by which enough samples can be obtained at very high altitudes. This study will form part of the UK contribution to the wider international studies relating to SO_2 in the atmosphere sponsored by bodies such as the OECD.

Smoke, grit and dust
Under the 1965 Clean Air Act dark smoke emissions—Ringelmann 2 or darker—are prohibited except for limited and carefully defined periods relating to cold light up of plant, soot blowing conditions, unforeseen plant failure *etc*. The Act also included provisions for the introduction by local authorities of smoke control areas. In these areas visible smoke emissions are prohibited

and this applies to all users, industrial and domestic. It is achieved either by the use of smokeless fuels or by ensuring that combustion equipment and standards for both domestic and industrial equipment more than comply with the requirements.

The 1968 Clean Air Act shows much greater concern with grit and dust emissions. In effect, the restrictions of the 1956 Act were extended to cover emissions of these types from almost all industrial and commercial premises. Oil firing techniques and developments have played an important role in ensuring minimum emissions and the atmosphere is progressively being cleared of smoke and dust particles. Where the provisions are fully implemented the benefits can be seen clearly.

Products of partial combustion

Petroleum fuels used for 'continuous' processes of domestic and industrial heating do not normally produce a significant amount of products of partial combustion. But the petrol engine operates under quite different conditions, on a cyclic basis. The combustion reaction has to take place in a small water cooled chamber with a rapidly fluctuating temperature and pressure and must be complete within a very small fraction of a second. Thus, it is not surprising that some products of partial combustion are produced. Table 3 gives their concentrations for various driving modes in an average car exhaust.

Unburnt hydrocarbons are known to play a key role in the formation of photochemical (Los Angeles) smog. They react with nitrogen oxides and ozone present in the atmosphere to produce a number of lachrymatory substances. Peroxyacyl nitrates, for example, have been detected in the air above Los Angeles.[11] There, the peculiar topographical situation, weather conditions and very high car population have made this form of pollution particularly severe. So far, its occurrence elsewhere is comparatively rare and the weather conditions that exist in the UK are such that there appears to be no real likelihood of it occurring here.

The toxic properties of carbon monoxide are well known so it is not surprising that attention should be focussed on its presence in exhaust emissions. On average, a petrol engine exhaust contains about 4 per cent CO by volume but this is usually quickly dispersed into the surrounding atmosphere. Surveys made in six towns in this country and confirmed by other studies abroad showed typical day to day concentrations at nose level of about 1 ppm and although higher peaks did occur 99 per cent of them were below 20 ppm.[5]

Carbon monoxide combines with the haemoglobin of the blood and can be measured in its combined form as carboxyhaemoglobin (COHb). Tests have shown that smoking is a much more potent source of carbon monoxide than ingestion of automobile exhausts. After exposure of both non-smokers and smokers to traffic fumes, non-smokers had on average about 1.2–1.9 per cent COHb while smokers had twice as much. The evidence so far is that COHb levels have to reach about 8 per cent before even minor effects, such as impaired ability to carry out mental tests, are manifested.[12] The Medical Research Council Unit engaged in these studies now proposes to extend its work to look at the possible synergistic effect of other pollutants present in

Table 4. Present and proposed US federal gasoline emission requirements.

Compound	1971	1975*
Exhaust hydrocarbons (g mile^{-1})	2.2	0.5
Exhaust carbon monoxide (g mile^{-1})	23	11
Exhaust NO$_x$ (g mile^{-1})	4·0†	0.9
Evaporative hydrocarbon emissions (g test^{-1})	6	2
Crankcase hydrocarbon	None allowed	None allowed
Particulates (g mile^{-1})	No requirement	0.1

† Operative in California only.   * In addition there will be a change in the standard test cycle used as the basis of assessment. This will have the effect of making the 1975 requirements even more severe.

exhaust emissions.

At least part of the carbon monoxide concentrations recorded in Los Angeles may have originated in the ocean, since scientists from the Max Planck Institute have found that about two-thirds of the carbon monoxide emitted to the atmosphere is produced in unknown chemical reactions taking place in sea water.

On this evidence there seems to be little need to take draconian steps in the UK to control exhaust emissions, whatever the attitude may be elsewhere. In the US, legislation is already in existence to control the quantity of exhaust emissions and the limits are to become even more stringent (Table 4). Already several major US car manufacturers are claiming that they are unable to meet the 1975 requirements by that date.[13] Whether or not this is so, reduction of exhaust emissions to the proposed limits can probably only be achieved by using an after-burner in the exhaust system which converts unburnt hydrocarbons and CO to CO_2. A thermal device has been developed but the general consensus at present is that a catalyst will need to be incorporated to promote the conversion. Unfortunately, the efficiency of these catalysts is seriously impaired by the lead compounds which are added to the petrol to improve its quality to meet the requirements of modern high compression-ratio engines. As far as the pollution aspects from lead are concerned probably more work has been done on this subject than on any other and there is no evidence that it is a significant hazard to health at its present concentrations. Nevertheless, the introduction of legislation on American lines will inevitably lead to the adoption of unleaded or very low leaded petrol; engines will have to be de-rated; and fuel consumption will increase by perhaps 15 per cent.

Unleaded petrol can certainly be made available, given a sensible time scale, but once again the cost will be high and will have to be passed on to the user. It is estimated[14] that it will add £5–£10 a year to the fuel bill of the average motorist and he will also have to pay £50–£100 extra for the exhaust emission equipment on his car and for its periodic checking and maintenance. On top of this will be the cost to the community for the enforcement of the legislation, which must involve the use of sophisticated roadside analytical equipment.

Diesel engine exhausts contain far less unburnt hydrocarbons and carbon monoxide than car exhausts (Table 5). If badly maintained they can be a prolific source of objectionable smoke and it could well be that much of the public reaction to exhaust emissions is provoked by such vehicles. Legislation already exists in the UK to deal with visible emissions whether from petrol or diesel vehicles and these could be eliminated if the law was enforced.

Nitrogen oxides

Above 1500 °C, atmospheric nitrogen and oxygen react to form nitric oxides, and although the reaction is reversible,

Table 5. Typical values for NO_x produced by various fuels (see Table 3 for motor vehicles).

Process	NO_x ppm Low	High
Oil-fired boiler	330	915
Gas-fired boiler	160	1140
Coal-fired boiler	650	1420

in practice there is little opportunity for this reversal to take place. The NO will react fairly slowly to produce NO_2 on discharge into the atmosphere. The reaction rate depends, amongst other things, on atmospheric temperature—a change of 10 °C can produce a 16 per cent difference in the amount of NO_2 formed.[15]

It is usual to refer to the mixed oxides as NO_x and Tables 3 and 5 show the amount produced by the combustion of various types of fuel. No special significance should be read into the maxima for the different fuels as the amount of NO_x produced is a function of the combustion temperature and oxygen availability. Thus the methods that are available to reduce NO_x are obviously limited. Petrol engines running either rich or lean produce less NO_x, but the former leads to significant losses of power whilst the latter increases the amount of partial combustion products. So far, the best results have come from re-cycling about 15 per cent of the exhaust gas, which has the effect of reducing peak flame temperature. Although catalytic methods of reducing NO_x are under development, they are still at a very early stage.

Measurements made in a crowded London street, where most of the NO_x would arise from internal combustion engines, have shown average mean hourly figures of 1.7 ppm NO and 0.2 ppm NO_2 measured at nose level.[12] Responsible medical opinion does not regard these concentrations as creating any significant health hazard.

Carbon dioxide

The main product of the combustion of any fossil fuel is CO_2. The total emission of carbon dioxide from natural sources, i.e. sources other than human activity, is estimated to be 10^{12} tons per annum whereas the amount of CO_2 resulting from human activity, i.e. combustion, is only 1.3 per cent of this annual emission. Carbon dioxide in the atmosphere allows sunlight to penetrate freely but partially absorbs outgoing radiation from the earth's surface. This should result in an increase in the mean surface temperature of the earth and, indeed, until 1940 measured temperatures appeared to parallel the increase of CO_2 in the atmosphere which in the past 50 years had risen from about 300 to about 330 ppm. Since 1940 the mean surface temperature has shown a slight decrease of about 0.1 per cent. This may be the result of an increase in particulate matter in the atmosphere, which reduces the amount of incoming sunlight and thus offsets the effect of CO_2. Whilst some of the particulate matter may come from the combustion of fossil fuels, seasonal variations are found which do not fit the general pattern. Volcanic activity has increased since 1940 and it is possible that volcanic dust suspended in the air is causing a reduction in the amount of sunlight reaching the earth.

If control of CO_2 emissions from combustion were considered necessary in the future it would pose technical problems and costs of a magnitude as yet unrealized.

The oil industry uses considerable amounts of its own products for the purposes of refining and distribution. In refining alone 5.6 Mton of petroleum fuels were used in 1969 and the industry is extremely conscious of its duty to do everything possible to minimize atmospheric pollution.[17] Most major oil companies provide a free technical service to users of their products: for petroleum fuels, this is largely concerned with their more efficient utilization and consequent reduction in pollution. They also conduct extensive research programmes for better techniques of fuel utilization.

A joint body, the British Technical Council of Motor and Petroleum Industries, is carrying out work on vehicle emissions; and the Institute of Petroleum gives support to many conservation activities, such as the National Society for Clean Air, and provides technical representation on many national and government committees engaged with pollution matters. In Europe the oil companies have set up an organization known as Stichting Concawe (*Con*servation of *C*lean *A*ir and *W*ater *W*estern *E*urope). This body disseminates information on environmental matters and provides technical advice. A recent report, *The impact of the oil industry on the environment*, is a valuable source book of data.[18]

Conclusions

About a half of the UK's present energy demands are met by petroleum fuels and without this contribution the high standard of living we enjoy could not possibly be sustained. The use of petroleum fuels for industry and home heating has resulted in an overall saving to the national economy both in terms of more efficient utilization and also in the benefits accruing from the reduction of the most serious forms of atmospheric pollution. The petrol engine has revolutionized our lives and it is entirely unrealistic to believe that any alternative power source, which is as flexible and economic, could replace it in the forseeable future.[14]

As a result of the Clean Air Acts, the air in this country is now a good deal cleaner than it was 20 years ago, and it is not surprising that attention has turned to other man-made additions to the atmosphere. Unfortunately, the **subject of atmospheric pollution is now acquiring political and emotive overtones and pronouncements are being made which are based on little other than pure speculation.**

179

The Royal College of Physicians has recently produced an authoritative report on the effect of air pollution on the health of the community.[19] Whilst this raises some questions which are still unanswered, its main recommendations are that the Clean Air Acts should be fully implemented in all parts of the country and more efficient methods should be used for the disposal of combustion products in order to avoid unnecessarily high local concentrations. District heating schemes by which a large central boiler house supplies hot water to the surrounding community is one of the ways of reducing atmospheric pollution which the report quotes and oil companies have pioneered these schemes in the UK.

More scientific effort is certainly required in some aspects of atmospheric pollution but with the resources that are now available it seems only too easy for projects to be started which add little or nothing to our existing store of knowledge but merely make further measurements to reinforce what we already know. The *Financial Times*[20] has drawn attention to some of the dangers inherent in the appearance of what it calls the conservation bandwagon. This seems to be borne out by the rapid proliferation of bodies set up to investigate pollution, and, in some cases, by the departure from the strict scientific canons of evidence which their deliberations display. Last year, 100 American scientists tried to return to first principles and spent a month examining the consequences of pollution in terms of changes of climate, ocean ecology and large terrestrial eco-systems.[21] Of the pollutants from petroleum fuels considered—CO_2, SO_2, nitrogen oxides and heavy metals— only CO_2 was regarded as a pollutant worthy of further consideration, and they thought it unlikely that there would be any direct climatic change in this century resulting from an increase in CO_2 in the atmosphere.

In the end, it is politicians and public opinion that decide how our environment should be improved and the priorities between the competing areas seeking attention. Improvements can certainly be made in the fields where petroleum products are used, but they will usually be subject to the law of diminishing returns[22] and add extra cost to the products involved. There are other forms of pollution, for instance of rivers, which many claim are of more immediate concern than atmospheric pollution. Chemists have an invaluable part to play here in helping government and public opinion to formulate a rational scale of priorities for environmental conservation, based on the known facts.

References
1. G. Whittingham, *3rd Int. Symp. on Combustion*, 1949, 433.
2. J. W. Laxton, *The mechanism of corrosion by fuel impurities*, 228. London: Butterworths, 1963.
3. A. Parker, in *The Clean Air Book for 1970-71*. London: National Society for Clean Air, 197.
4. R. Farnsworth and H. Rowling, *Proc. Inst. Petrol. Summer Meeting*, 1968.
5. A. W. Pearce and C. S. Windebank, Clean Air Congress, Southport, 1970.
6. A. M. Squires, *Chem. Engr, Lond.*, 1967, **74**, 133.
7. *Chem. Proc. Engng*, 1970, **8**, 5.
8. P. J. Lawther, Clean Air Congress, Eastbourne, 1969.
9. K. J. Marsh and V. R.Withers, *Atmosph. Envir.*, 1969, **3**, 281.
10. P. C. Blocker, *J. Inst. Petrol.*, 1970, **56**, 71.
11. E. R. Stephens, *Adv. envir. Sci. Technol.*, 1969, **1**, 121.
12. P. J. Lawther and L. R. Commings, Clean Air Congress, Southport, 1970.
13. *Chem. engng News*, 22 June 1970, 15.
14. Lord Rothschild, *Petrol and pollution*. The Royal Society Technology Lecture, 1970.
15. E. A. Schuck and E. R. Stephens, *Adv. envir. Sci. Technol.*, 1969, **1**, 75.
16. D. H. Barnhart and E. K. Diehl, *J. Air Pollut. Control Ass.*, 1960, **10**, 399.
17. M. Benger, *Proc. Inst. Petrol. Summer Meeting*, 1965.
18. W. C. Hopper and B. Rayzacher, *The impact of the oil industry on the environment*. The Hague: Stichting Concawe, 1970.
19. *Air pollution and health*—a report for the Royal College of Physicians. London: Pitman, 1970.
20. D. Fishlock, *Financial Times*, 30 July 1970.
21. *Bull. atom. Scient.*, 1970, **26**, 25.
22. C. Sinclair, *The management of hazard in industrial innovation*. London: Industrial Educational and Research Foundation, 1970.

AUTHOR INDEX

Alarie, Yves, 47
Allison, A.C., 101

Battigelli, Mario C., 93
Behrman, Richard E., 61
Busey, William M., 47

Clausen, Jack L., 26
Cohen, Charles A., 26
Cole, Homer, 93
Cota, Harold M., 165
Cotton, Raymond D., 118

DuBois, Arthur B., 72

Fisher, David E., 61
Fraser, David A., 93

Gay, Michael L., 97
Goldstein, E., 10

Hudson, Arnold R., 26

Knelson, John H., 26
Krumm, Alex A., 47

MacFarland, Harold N., 47

Neal, Jack, 78

Parkinson, G.S., 173
Paton, John, 61

Rigdon, R.H., 78

Scorer, R.S., 122
Shiffman, Morris A., 154

Ulrich, Charles E., 47

Winkelstein, Warren, Jr., 97

KEY-WORD TITLE INDEX